Advanced FPGA
Design

BICENTENNIAL
1807
🟫WILEY
2007
BICENTENNIAL

The Wiley Bicentennial—Knowledge for Generations

*E*ach generation has its unique needs and aspirations. When Charles Wiley first opened his small printing shop in lower Manhattan in 1807, it was a generation of boundless potential searching for an identity. And we were there, helping to define a new American literary tradition. Over half a century later, in the midst of the Second Industrial Revolution, it was a generation focused on building the future. Once again, we were there, supplying the critical scientific, technical, and engineering knowledge that helped frame the world. Throughout the 20th Century, and into the new millennium, nations began to reach out beyond their own borders and a new international community was born. Wiley was there, expanding its operations around the world to enable a global exchange of ideas, opinions, and know-how.

For 200 years, Wiley has been an integral part of each generation's journey, enabling the flow of information and understanding necessary to meet their needs and fulfill their aspirations. Today, bold new technologies are changing the way we live and learn. Wiley will be there, providing you the must-have knowledge you need to imagine new worlds, new possibilities, and new opportunities.

Generations come and go, but you can always count on Wiley to provide you the knowledge you need, when and where you need it!

WILLIAM J. PESCE
PRESIDENT AND CHIEF EXECUTIVE OFFICER

PETER BOOTH WILEY
CHAIRMAN OF THE BOARD

Advanced FPGA Design
Architecture, Implementation, and Optimization

Steve Kilts
Spectrum Design Solutions
Minneapolis, Minnesota

IEEE

The Institute of Electrical and Electronics Engineers, Inc., New York

BICENTENNIAL

1807

WILEY

2007

BICENTENNIAL

WILEY-INTERSCIENCE
A JOHN WILEY & SONS, INC., PUBLICATION

Library of Congress Cataloging-in-Publication Data

Kilts, Steve, 1978-
    Advanced FPGA design: Architecture, Implementation, and Optimization/
by Steve Kilts.
        p. cm.
    Includes index.
    ISBN   978-0-470-05437-6 (cloth)
    1. Field programmable gate arrays- -Design and construction.
I. Title.
    TK7895.G36K55 2007
    621.39′5- -dc22
2006033573

10   9

*To my wife, Teri, who felt that the
subject matter was rather dry*

Flowchart of Contents

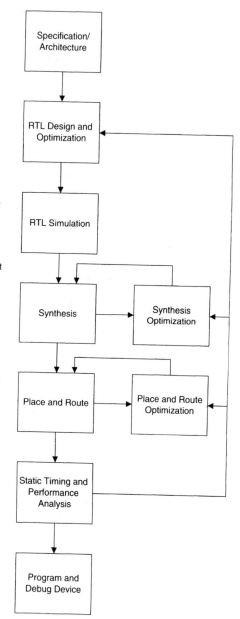

Contents

Preface

In the design-consulting business, I have been exposed to countless FPGA (Field Programmable Gate Array) designs, methodologies, and design techniques. Whether my client is on the Fortune 100 list or is just a start-up company, they will inevitably do some things right and many things wrong. After having been exposed to a wide variety of designs in a wide range of industries, I began developing my own arsenal of techniques and heuristics from the combined knowledge of these experiences. When mentoring new FPGA design engineers, I draw my suggestions and recommendations from this experience. Up until now, many of these recommendations have referenced specific white papers and application notes (appnotes) that discuss specific practical aspects of FPGA design. The purpose of this book is to condense years of experience spread across numerous companies and teams of engineers, as well as much of the wisdom gathered from technology-specific white papers and appnotes, into a single book that can be used to refine a designer's knowledge and aid in becoming an advanced FPGA designer.

There are a number of books on FPGA design, but few of these truly address advanced real-world topics in detail. This book attempts to cut out the fat of unnecessary theory, speculation on future technologies, and the details of outdated technologies. It is written in a terse, concise format that addresses the various topics without wasting the reader's time. Many sections in this book assume that certain fundamentals are understood, and for the sake of brevity, background information and/or theoretical frameworks are not always covered in detail. Instead, this book covers in-depth topics that have been encountered in real-world designs. In some ways, this book replaces a limited amount of industry experience and access to an experienced mentor and will hopefully prevent the reader from learning a few things the hard way. It is the advanced, practical approach that makes this book unique.

One thing to note about this book is that it will not flow from cover to cover like a novel. For a set of advanced topics that are not intrinsically tied to one another, this type of flow is impossible without blatantly filling it with fluff. Instead, to organize this book, I have ordered the chapters in such a way that they follow a typical design flow. The first chapters discuss architecture, then simulation, then synthesis, then floorplanning, and so on. This is illustrated in the Flowchart of Contents provided at the beginning of the book. To provide accessibility for future reference, the chapters are listed side-by-side with the relevant block in the flow diagram.

The remaining chapters in this book are heavy with examples. For brevity, I have selected Verilog as the default HDL (Hardware Description Language), Xilinx as the default FPGA vendor, and Synplicity as the default synthesis and floorplanning tool. Most of the topics covered in this book can easily be mapped to VHDL, Altera, Mentor Graphics, and so forth, but to include all of these for completeness would only serve to cloud the important points. Even if the reader of this book uses these other technologies, this book will still deliver its value. If you have any feedback, good or bad, feel free to email me at steve.kilts@spectrumdsi.com

STEVE KILTS

Minneapolis, Minnesota
March 2007

Acknowledgments

During the course of my career, I have had the privilege to work with many excellent digital design engineers. My exposure to these talented engineers began at Medtronic and continued over the years through my work as a consultant for companies such as Honeywell, Guidant, Teradyne, Telex, Unisys, AMD, ADC, and a number of smaller/start-up companies involved with a wide variety of FPGA applications. I also owe much of my knowledge to the appnotes and white papers published by the major FPGA vendors. These resources contain invaluable real-world heuristics that are not included in a standard engineering curriculum.

Specific to this book, I owe a great deal to Xilinx and Synplicity, both of which provided the FPGA design tools used throughout the book, as well as a number of key reviewers. Reviewers of note also include Peter Calabrese of Synplicity, Cliff Cummins of Sunburst Design, Pete Danile of Synplicity, Anders Enggaard of Axcon, Mike Fette of Spectrum Design Solutions, Philip Freidin of Fliptronics, Paul Fuchs of NuHorizons, Don Hodapp of Xilinx, Ashok Kulkarni of Synplicity, Rod Landers of Spectrum Design Solutions, Ryan Link of Logic, Dave Matthews of Verein, Lance Roman of Roman-Jones, B. Joshua Rosen of Polybus, Gary Stevens of iSine, Jim Torgerson, and Larry Weegman of Xilinx.

S.K.

Chapter 1

Architecting Speed

Sophisticated tool optimizations are often not good enough to meet most design constraints if an arbitrary coding style is used. This chapter discusses the first of three primary physical characteristics of a digital design: speed. This chapter also discusses methods for architectural optimization in an FPGA.

There are three primary definitions of speed depending on the context of the problem: throughput, latency, and timing. In the context of processing data in an FPGA, throughput refers to the amount of data that is processed per clock cycle. A common metric for throughput is bits per second. Latency refers to the time between data input and processed data output. The typical metric for latency will be time or clock cycles. Timing refers to the logic delays between sequential elements. When we say a design does not "meet timing," we mean that the delay of the critical path, that is, the largest delay between flip-flops (composed of combinatorial delay, clk-to-out delay, routing delay, setup timing, clock skew, and so on) is greater than the target clock period. The standard metrics for timing are clock period and frequency.

During the course of this chapter, we will discuss the following topics in detail:

- High-throughput architectures for maximizing the number of bits per second that can be processed by the design.
- Low-latency architectures for minimizing the delay from the input of a module to the output.
- Timing optimizations to reduce the combinatorial delay of the critical path.

  Adding register layers to divide combinatorial logic structures.

  Parallel structures for separating sequentially executed operations into parallel operations.

  Flattening logic structures specific to priority encoded signals.

  Register balancing to redistribute combinatorial logic around pipelined registers.

  Reordering paths to divert operations in a critical path to a noncritical path.

Advanced FPGA Design. By Steve Kilts
Copyright © 2007 John Wiley & Sons, Inc.

1.1   HIGH THROUGHPUT

A high-throughput design is one that is concerned with the steady-state data rate but less concerned about the time any specific piece of data requires to propagate through the design (latency). The idea with a high-throughput design is the same idea Ford came up with to manufacture automobiles in great quantities: an assembly line. In the world of digital design where data is processed, we refer to this under a more abstract term: pipeline.

A pipelined design conceptually works very similar to an assembly line in that the raw material or data input enters the front end, is passed through various stages of manipulation and processing, and then exits as a finished product or data output. The beauty of a pipelined design is that new data can begin processing before the prior data has finished, much like cars are processed on an assembly line. Pipelines are used in nearly all very-high-performance devices, and the variety of specific architectures is unlimited. Examples include CPU instruction sets, network protocol stacks, encryption engines, and so on.

From an algorithmic perspective, an important concept in a pipelined design is that of "unrolling the loop." As an example, consider the following piece of code that would most likely be used in a software implementation for finding the third power of X. Note that the term "software" here refers to code that is targeted at a set of procedural instructions that will be executed on a microprocessor.

```
XPower = 1;
for (i=0;i < 3; i++)
  XPower = X * XPower;
```

Note that the above code is an iterative algorithm. The same variables and addresses are accessed until the computation is complete. There is no use for parallelism because a microprocessor only executes one instruction at a time (for the purpose of argument, just consider a single core processor). A similar implementation can be created in hardware. Consider the following Verilog implementation of the same algorithm (output scaling not considered):

```
module power3(
  output [7:0] XPower,
  output       finished,
  input  [7:0] X,
  input        clk, start); // the duration of start is a
                            //                  single clock
  reg    [7:0] ncount;
  reg    [7:0] XPower;

  assign finished = (ncount == 0);

  always@(posedge clk)
    if(start) begin
      XPower <= X;
      ncount <= 2;
    end
```

```
        else if(!finished) begin
            ncount <= ncount - 1;
            XPower <= XPower * X;
        end
endmodule
```

In the above example, the same register and computational resources are reused until the computation is finished as shown in Figure 1.1.

With this type of iterative implementation, no new computations can begin until the previous computation has completed. This iterative scheme is very similar to a software implementation. Also note that certain handshaking signals are required to indicate the beginning and completion of a computation. An external module must also use the handshaking to pass new data to the module and receive a completed calculation. The performance of this implementation is

Throughput = 8/3, or 2.7 bits/clock

Latency = 3 clocks

Timing = One multiplier delay in the critical path

Contrast this with a pipelined version of the same algorithm:

```
module power3(
    output reg [7:0] XPower,
    input            clk,
    input      [7:0] X
    );
    reg        [7:0] XPower1, XPower2;
    reg        [7:0] X1, X2;
    always @(posedge clk) begin
        // Pipeline stage 1
        X1       <= X;
        XPower1 <= X;

        // Pipeline stage 2
        X2       <= X1;
        XPower2 <= XPower1 * X1;

        // Pipeline stage 3
        XPower <= XPower2 * X2;
    end
endmodule
```

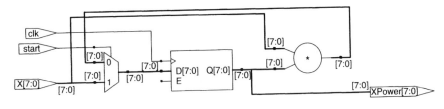

Figure 1.1   Iterative implementation.

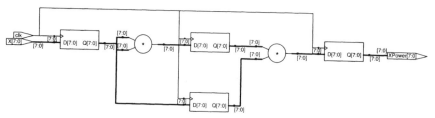

Figure 1.2 Pipelined implementation.

In the above implementation, the value of X is passed to both pipeline stages where independent resources compute the corresponding multiply operation. Note that while X is being used to calculate the final power of 3 in the second pipeline stage, the next value of X can be sent to the first pipeline stage as shown in Figure 1.2.

Both the final calculation of X^3 (XPower3 resources) and the first calculation of the next value of X (XPower2 resources) occur simultaneously. The performance of this design is

Throughput $= 8/1$, or 8 bits/clock

Latency $= 3$ clocks

Timing $=$ One multiplier delay in the critical path

The throughput performance increased by a factor of 3 over the iterative implementation. In general, if an algorithm requiring n iterative loops is "unrolled," the pipelined implementation will exhibit a throughput performance increase of a factor of n. There was no penalty in terms of latency as the pipelined implementation still required 3 clocks to propagate the final computation. Likewise, there was no timing penalty as the critical path still contained only one multiplier.

Unrolling an iterative loop increases throughput.

The penalty to pay for unrolling loops such as this is an increase in area. The iterative implementation required a single register and multiplier (along with some control logic not shown in the diagram), whereas the pipelined implementation required a separate register for both X and XPower and a separate multiplier for every pipeline stage. Optimizations for area are discussed in the Chapter 2.

The penalty for unrolling an iterative loop is a proportional increase in area.

1.2 LOW LATENCY

A low-latency design is one that passes the data from the input to the output as quickly as possible by minimizing the intermediate processing delays. Oftentimes, a low-latency design will require parallelisms, removal of pipelining, and logical short cuts that may reduce the throughput or the max clock speed in a design.

Referring back to our power-of-3 example, there is no obvious latency optimization to be made to the iterative implementation as each successive multiply operation must be registered for the next operation. The pipelined implementation, however, has a clear path to reducing latency. Note that at each pipeline stage, the product of each multiply must wait until the next clock edge before it is propagated to the next stage. By removing the pipeline registers, we can minimize the input to output timing:

```
module power3(
    output [7:0] XPower,
    input  [7:0] X
    );
    reg    [7:0] XPower1, XPower2;
    reg    [7:0] X1, X2;

    assign XPower = XPower2 * X2;

    always @* begin
      X1      = X;
      XPower1 = X;
    end

    always @* begin
      X2      = X1;
      XPower2 = XPower1*X1;
    end
endmodule
```

In the above example, the registers were stripped out of the pipeline. Each stage is a combinatorial expression of the previous as shown in Figure 1.3.

The performance of this design is

Throughput = 8 bits/clock (assuming one new input per clock)

Latency = Between one and two multiplier delays, 0 clocks

Timing = Two multiplier delays in the critical path

By removing the pipeline registers, we have reduced the latency of this design below a single clock cycle.

Latency can be reduced by removing pipeline registers.

The penalty is clearly in the timing. Previous implementations could theoretically run the system clock period close to the delay of a single multiplier, but in the

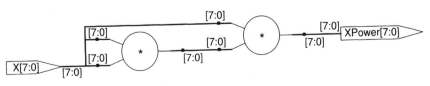

Figure 1.3   Low-latency implementation.

low-latency implementation, the clock period must be at least two multiplier delays (depending on the implementation) plus any external logic in the critical path.

The penalty for removing pipeline registers is an increase in combinatorial delay between registers.

1.3   TIMING

Timing refers to the clock speed of a design. The maximum delay between any two sequential elements in a design will determine the max clock speed. The idea of clock speed exists on a lower level of abstraction than the speed/area trade-offs discussed elsewhere in this chapter as clock speed in general is not directly related to these topologies, although trade-offs within these architectures will certainly have an impact on timing. For example, one cannot know whether a pipelined topology will run faster than an iterative without knowing the details of the implementation. The maximum speed, or maximum frequency, can be defined according to the straightforward and well-known maximum-frequency equation (ignoring clock-to-clock jitter):

Equation 1.1 Maximum Frequency

$$F_{max} = \frac{1}{T_{clk-q} + T_{logic} + T_{routing} + T_{setup} - T_{skew}} \qquad (1.1)$$

where F_{max} is maximum allowable frequency for clock; T_{clk-q} is time from clock arrival until data arrives at Q; T_{logic} is propagation delay through logic between flip-flops; $T_{routing}$ is routing delay between flip-flops; T_{setup} is minimum time data must arrive at D before the next rising edge of clock (setup time); and T_{skew} is propagation delay of clock between the launch flip-flop and the capture flip-flop.

The next sections describes various methods and trade-offs required to improve timing performance.

1.3.1   Add Register Layers

The first strategy for architectural timing improvements is to add intermediate layers of registers to the critical path. This technique should be used in highly pipelined designs where an additional clock cycle latency does not violate the design specifications, and the overall functionality will not be affected by the further addition of registers.

For instance, assume the architecture for the following FIR (Finite Impulse Response) implementation does not meet timing:

```
module fir(
   output [7:0] Y,
   input  [7:0] A, B, C, X,
   input        clk,
```

```
input          validsample);
reg    [7:0] X1, X2, Y;

always @(posedge clk)
  if(validsample) begin
    X1 <= X;
    X2 <= X1;
    Y  <= A* X+B* X1+C* X2;
  end
endmodule
```

Architecturally, all multiply/add operations occur in one clock cycle as shown in Figure 1.4.

In other words, the critical path of one multiplier and one adder is greater than the minimum clock period requirement. Assuming the latency requirement is not fixed at 1 clock, we can further pipeline this design by adding extra registers intermediate to the multipliers. The first layer is easy: just add a pipeline layer between the multipliers and the adder:

```
module fir(
  output [7:0] Y,
  input  [7:0] A, B, C, X,
  input        clk,
  input        validsample);
  reg    [7:0] X1, X2, Y;
  reg    [7:0] prod1, prod2, prod3;

  always @ (posedge clk) begin
    if(validsample) begin
      X1    <= X;
      X2    <= X1;
      prod1 <= A * X;
      prod2 <= B * X1;
      prod3 <= C * X2;
    end
    Y <= prod1 + prod2 + prod3;
  end
endmodule
```

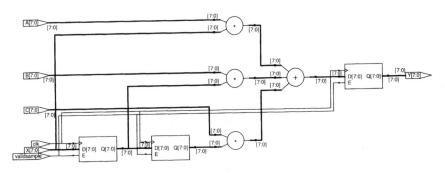

Figure 1.4 MAC with long path.

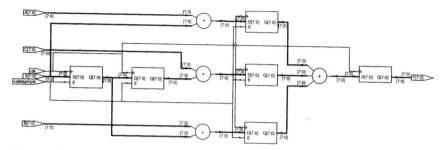

Figure 1.5 Pipeline registers added.

In the above example, the adder was separated from the multipliers with a pipeline stage as shown in Figure 1.5.

Multipliers are good candidates for pipelining because the calculations can easily be broken up into stages. Additional pipelining is possible by breaking the multipliers and adders up into stages that can be individually registered.

> Adding register layers improves timing by dividing the critical path into two paths of smaller delay.

Various implementations of these functions are covered in other chapters, but once the architecture has been broken up into stages, additional pipelining is as straightforward as the above example.

1.3.2 Parallel Structures

The second strategy for architectural timing improvements is to reorganize the critical path such that logic structures are implemented in parallel. This technique should be used whenever a function that currently evaluates through a serial string of logic can be broken up and evaluated in parallel. For instance, assume that the standard pipelined power-of-3 design discussed in previous sections does not meet timing. To create parallel structures, we can break the multipliers into independent operations and then recombine them. For instance, an 8-bit binary multiplier can be represented by nibbles A and B:

$$X = \{A, B\},$$

where A is the most significant nibble and B is the least significant.

Because the multiplicand is equal to the multiplier in our power-of-3 example, the multiply operation can be reorganized as follows:

$$X * X = \{A, B\} * \{A, B\} = \{(A * A), (2 * A * B), (B * B)\};$$

This reduces our problem to a series of 4-bit multiplications and then recombining the products. This can be implemented with the following module:

```
module power3 (
    output [7:0] XPower,
```

```
input   [7:0] X,
input         clk);
reg     [7:0] XPower1;
// partial product registers
reg     [3:0] XPower2_ppAA, XPower2_ppAB, XPower2_ppBB;
reg     [3:0] XPower3_ppAA, XPower3_ppAB, XPower3_ppBB;
reg     [7:0] X1, X2;
wire    [7:0] XPower2;

// nibbles for partial products (A is MS nibble, B is LS
                                                nibble)
wire    [3:0] XPower1_A = XPower1[7:4];
wire    [3:0] XPower1_B = XPower1[3:0];
wire    [3:0] X1_A      = X1[7:4];
wire    [3:0] X1_B      = X1[3:0];
wire    [3:0] XPower2_A = XPower2[7:4];
wire    [3:0] XPower2_B = XPower2[3:0];
wire    [3:0] X2_A      = X2[7:4];
wire    [3:0] X2_B      = X2[3:0];

// assemble partial products
assign XPower2        = (XPower2_ppAA << 8)+
                        (2*XPower2_ppAB << 4)+
                         XPower2_ppBB;
assign XPower         = (XPower3_ppAA << 8)+
                        (2*XPower3_ppAB << 4)+
                         XPower3_ppBB;

always @(posedge clk) begin
  // Pipeline stage 1
  X1           <= X;
  XPower1      <= X;

  // Pipeline stage 2
  X2           <= X1;
  // create partial products
  XPower2_ppAA <= XPower1_A * X1_A;
  XPower2_ppAB <= XPower1_A * X1_B;
  XPower2_ppBB <= XPower1_B * X1_B;

  // Pipeline stage 3
  // create partial products
  XPower3_ppAA <= XPower2_A * X2_A;
  XPower3_ppAB <= XPower2_A * X2_B;
  XPower3_ppBB <= XPower2_B * X2_B;
  end
endmodule
```

This design does not take into consideration any overflow issues, but it serves to illustrate the point. The multiplier was broken down into smaller functions that could be operated on independently as shown in Figure 1.6.

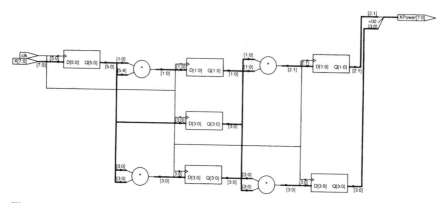

Figure 1.6 Multiplier with separated stages.

By breaking the multiply operation down into smaller operations that can execute in parallel, the maximum delay is reduced to the longest delay through any of the substructures.

Separating a logic function into a number of smaller functions that can be evaluated in parallel reduces the path delay to the longest of the substuctures.

1.3.3 Flatten Logic Structures

The third strategy for architectural timing improvements is to flatten logic structures. This is closely related to the idea of parallel structures defined in the previous section but applies specifically to logic that is chained due to priority encoding. Typically, synthesis and layout tools are smart enough to duplicate logic to reduce fanout, but they are not smart enough to break up logic structures that are coded in a serial fashion, nor do they have enough information relating to the priority requirements of the design. For instance, consider the following control signals coming from an address decode that are used to write four registers:

```
module regwrite(
  output reg [3:0] rout,
  input            clk, in,
  input      [3:0] ctrl);

  always @(posedge clk)
    if(ctrl[0])       rout[0] <= in;
    else if(ctrl[1]) rout[1] <= in;
    else if(ctrl[2]) rout[2] <= in;
    else if(ctrl[3]) rout[3] <= in;
endmodule
```

In the above example, each of the control signals are coded with a priority relative to the other control signals. This type of priority encoding is implemented as shown in Figure 1.7.

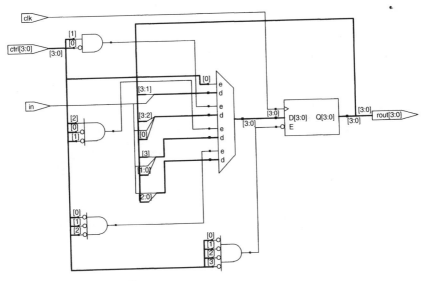

Figure 1.7   Priority encoding.

If the control lines are strobes from an address decoder in another module, then each strobe is mutually exclusive to the others as they all represent a unique address. However, here we have coded this as if it were a priority decision. Due to the nature of the control signals, the above code will operate exactly as if it were coded in a parallel fashion, but it is unlikely the synthesis tool will be smart enough to recognize that, particularly if the address decode takes place behind another layer of registers.

To remove the priority and thereby flatten the logic, we can code this module as shown below:

```
module regwrite(
    output reg [3:0] rout,
    input            clk, in,
    input      [3:0] ctrl);

    always @(posedge clk) begin
        if(ctrl[0]) rout[0] <= in;
        if(ctrl[1]) rout[1] <= in;
        if(ctrl[2]) rout[2] <= in;
        if(ctrl[3]) rout[3] <= in;
    end
endmodule
```

As can be seen in the gate-level implementation, no priority logic is used as shown in Figure 1.8. Each of the control signals acts independently and controls its corresponding rout bits independently.

By removing priority encodings where they are not needed, the logic structure is flattened and the path delay is reduced.

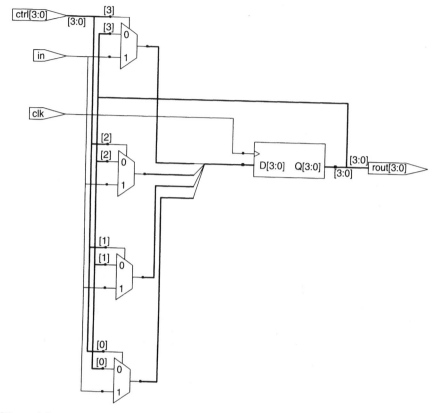

Figure 1.8   No priority encoding.

1.3.4 Register Balancing

The fourth strategy is called register balancing. Conceptually, the idea is to redistribute logic evenly between registers to minimize the worst-case delay between any two registers. This technique should be used whenever logic is highly imbalanced between the critical path and an adjacent path. Because the clock speed is limited by only the worst-case path, it may only take one small change to successfully rebalance the critical logic.

Many synthesis tools also have an optimization called register balancing. This feature will essentially recognize specific structures and reposition registers around logic in a predetermined fashion. This can be useful for common structures such as large multipliers but is limited and will not change your logic nor recognize custom functionality. Depending on the technology, it may require more expensive synthesis tools to implement. Thus, it is very important to understand this concept and have the ability to redistribute logic in custom logic structures.

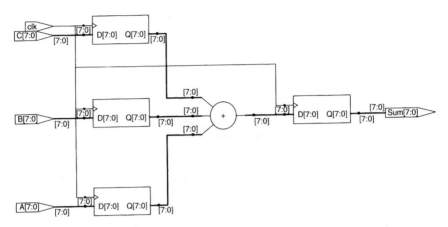

Figure 1.9   Registered adder.

Note the following code for an adder that adds three 8-bit inputs:

```
module adder(
    output reg [7:0] Sum,
    input      [7:0] A, B, C,
    input            clk);
    reg        [7:0] rA, rB, rC;

    always @(posedge clk) begin
        rA   <= A;
        rB   <= B;
        rC   <= C;
        Sum  <= rA + rB + rC;
    end
endmodule
```

The first register stage consists of rA, rB, and rC, and the second stage consists of Sum. The logic between stages 1 and 2 is the adder for all inputs, whereas the logic between the input and the first register stage contains no logic (assume the outputs feeding this module are registered) as shown in Figure 1.9.

If the critical path is defined through the adder, some of the logic in the critical path can be moved back a stage, thereby balancing the logic load between the two register stages. Consider the following modification where one of the add operations is moved back a stage:

```
module adder(
    output reg [7:0] Sum,
    input      [7:0] A, B, C,
    input            clk);
    reg        [7:0] rABSum, rC;
```

```
always @(posedge clk) begin
  rABSum <= A + B;
  rC     <= C;
  Sum    <= rABSum + rC;
end
endmodule
```

We have now moved one of the add operations back one stage between the input and the first register stage. This balances the logic between the pipeline stages and reduces the critical path as shown in Figure 1.10.

Register balancing improves timing by moving combinatorial logic from the critical path to an adjacent path.

1.3.5 Reorder Paths

The fifth strategy is to reorder the paths in the data flow to minimize the critical path. This technique should be used whenever multiple paths combine with the critical path, and the combined path can be reordered such that the critical path can be moved closer to the destination register. With this strategy, we will only be concerned with the logic paths between any given set of registers. Consider the following module:

```
module randomlogic(
  output reg [7:0] Out,
  input      [7:0] A, B, C,
  input            clk,
  input            Cond1, Cond2);
  always @(posedge clk)
    if(Cond1)
    Out <= A;
    else if(Cond2 && (C < 8))
    Out <= B;
    else
    Out <= C;
endmodule
```

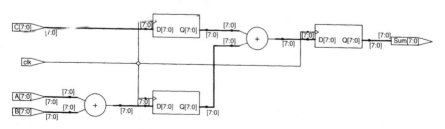

Figure 1.10   Registers balanced.

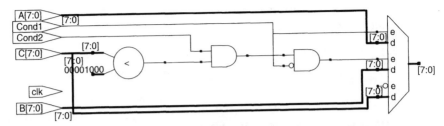

Figure 1.11   Long critical path.

In this case, let us assume the critical path is between C and Out and consists of a comparator in series with two gates before reaching the decision mux. This is shown in Figure 1.11. Assuming the conditions are not mutually exclusive, we can modify the code to reorder the long delay of the comparitor:

```
module randomlogic(
  output reg  [7:0] Out,
  input       [7:0] A, B, C,
  input             clk,
  input             Cond1, Cond2);
  wire CondB = (Cond2 & !Cond1);

  always @(posedge clk)
    if(CondB && (C < 8))
      Out <= B;
    else if(Cond1)
      Out <= A;
    else
      Out <= C;
endmodule
```

By reorganizing the code, we have moved one of the gates out of the critical path in series with the comparator as shown in Figure 1.12. Thus, by paying careful

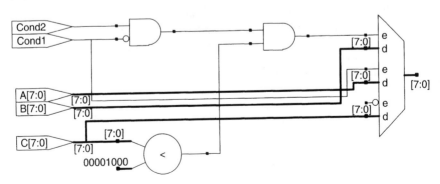

Figure 1.12   Logic reordered to reduce critical path.

attention to exactly how a particular function is coded, we can have a direct impact on timing performance.

Timing can be improved by reordering paths that are combined with the critical path in such a way that some of the critical path logic is placed closer to the destination register.

1.4 SUMMARY OF KEY POINTS

- A high-throughput architecture is one that maximizes the number of bits per second that can be processed by a design.
- Unrolling an iterative loop increases throughput.
- The penalty for unrolling an iterative loop is a proportional increase in area.
- A low-latency architecture is one that minimizes the delay from the input of a module to the output.
- Latency can be reduced by removing pipeline registers.
- The penalty for removing pipeline registers is an increase in combinatorial delay between registers.
- Timing refers to the clock speed of a design. A design meets timing when the maximum delay between any two sequential elements is smaller than the minimum clock period.
- Adding register layers improves timing by dividing the critical path into two paths of smaller delay.
- Separating a logic function into a number of smaller functions that can be evaluated in parallel reduces the path delay to the longest of the substructures.
- By removing priority encodings where they are not needed, the logic structure is flattened, and the path delay is reduced.
- Register balancing improves timing by moving combinatorial logic from the critical path to an adjacent path.
- Timing can be improved by reordering paths that are combined with the critical path in such a way that some of the critical path logic is placed closer to the destination register.

Chapter 2

Architecting Area

This chapter discusses the second of three primary physical characteristics of a digital design: area. Here we also discuss methods for architectural area optimization in an FPGA.

We will discuss area reduction based on choosing the correct topology. Topology refers to the higher-level organization of the design and is not device specific. Circuit-level reduction as performed by the synthesis and layout tools refers to the minimization of the number of gates in a subset of the design and may be device specific.

A topology that targets area is one that reuses the logic resources to the greatest extent possible, often at the expense of throughput (speed). Very often this requires a recursive data flow, where the output of one stage is fed back to the input for similar processing. This can be a simple loop that flows naturally with the algorithm or it may be that the logic reuse is complex and requires special controls. This section describes both techniques and describes the necessary consequences in terms of performance penalties.

During the course of this chapter, we will discuss the following topics in detail:

- Rolling up the pipeline to reuse logic resources in different stages of a computation.
- Controls to manage the reuse of logic when a natural flow does not exist.
- Sharing logic resources between different functional operations.
- The impact of reset on area optimization.
  Impact of FPGA resources that lack reset capability.
  Impact of FPGA resources that lack set capability.
  Impact of FPGA resources that lack asynchronous reset capability.
  Impact of RAM reset.
  Optimization using set/reset pins for logic implementation.

Advanced FPGA Design. By Steve Kilts
Copyright © 2007 John Wiley & Sons, Inc.

2.1 ROLLING UP THE PIPELINE

The method of "rolling up the pipeline" is the opposite operation to that described in the previous chapter to improve throughput by "unrolling the loop" to achieve maximum performance. When we unrolled the loop to create a pipeline, we also increased the area by requiring more resources to hold intermediate values and replicating computational structures that needed to run in parallel. Conversely, when we want to minimize the area of a design, we must perform these operations in reverse; that is, roll up the pipeline so that logic resources can be reused. Thus, this method should be used when optimizing highly pipelined designs with duplicate logic in the pipeline stages.

> Rolling up the pipeline can optimize the area of pipelined designs with duplicated logic in the pipeline stages.

Consider the example of a fixed-point fractional multiplier. In this example, A is represented in normal integer format with the fixed point just to the right of the LSB, whereas the input B has a fixed point just to the left of the MSB. In other words, B scales A from 0 to 1.

```
module mult8(
  output  [7:0]   product,
  input   [7:0]   A,
  input   [7:0]   B,
  input           clk);
  reg     [15:0]  prod16;

  assign product = prod16[15:8];

  always @(posedge clk)
    prod16 <= A * B;

endmodule
```

With this implementation, a new product is generated on every clock. There isn't an obvious pipeline in this design as far as distinct sets of registers, but note that the multiplier itself is a fairly long chain of logic that is easily pipelined by adding intermediate register layers. It is this multiplier that we wish to "roll up." We will roll this up by performing the multiply with a series of shift and add operations as follows:

```
module mult8(
  output          done,
  output reg [7:0]  product,
  input   [7:0]   A,
  input   [7:0]   B,
  input           clk,
  input           start);
  reg     [4:0]   multcounter; // counter for number of
                               //     shift/adds
```

```
reg        [7:0] shiftB; // shift register for B
reg        [7:0] shiftA; // shift register for A

wire adden; // enable addition

assign adden = shiftB[7] & !done;
assign done = multcounter[3];

always @(posedge clk) begin
  // increment   multiply counter for shift/add ops
  if(start)       multcounter <= 0;
  else if(!done) multcounter <= multcounter + 1;

  // shift register for B
  if(start) shiftB <= B;
  else shiftB[7:0] <= {shiftB[6:0], 1'b0};

  // shift register for A
  if(start) shiftA <= A;
  else shiftA[7:0] <= {shiftA[7], shiftA[7:1]};

  // calculate multiplication
  if(start)       product <= 0;
  else if(adden) product <= product + shiftA;
end
endmodule
```

The multiplier is thus architected with an accumulator that adds a shifted version of A depending on the bits of B as shown in Figure 2.1. Thus, we completely eliminate the logic tree necessary to generate a multiply within a single clock and replace it with a few shift registers and an adder. This is a very compact form of a multiplier but will now require 8 clocks to complete a multiplication. Also note that no special controls were necessary to sequence through this multiply operation. We simply relied on a counter to tell us when to stop the shift and add operations. The next section describes situations where this control is not so trivial.

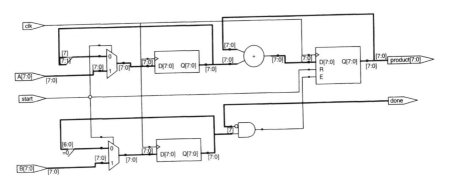

Figure 2.1   Shift/add multiplier.

2.2 CONTROL-BASED LOGIC REUSE

Sharing logic resources oftentimes requires special control circuitry to determine which elements are input to the particular structure. In the previous section, we described a multiplier that simply shifted the bits of each register, where each register was always dedicated to a particular input of the running adder. This had a natural data flow that lent itself well to logic reuse. In other applications, there are often more complex variations to the input of a resource, and certain controls may be necessary to reuse the logic.

> Controls can be used to direct the reuse of logic when the shared logic is larger than the control logic.

To determine this variation, a state machine may be required as an additional input to the logic.

Consider the following example of a low-pass FIR filter represented by the equation:

$$Y = coeffA * X[0] + coeffB * X[1] + coeffC * X[2]$$

```
module lowpassfir(
output reg [7:0] filtout,
output reg       done,
input            clk,
input      [7:0] datain, // X[0]
input            datavalid, // X[0] is valid
input      [7:0] coeffA, coeffB; coeffC); // coeffs for
                                          low pass
                                          filter
// define input/output samples
reg        [7:0] X0, X1, X2;
reg              multdonedelay;
reg              multstart; // signal to multiplier to
                               begin computation
reg        [7:0] multdat;
reg        [7:0] multcoeff; // the registers that are
                               multiplied together
reg        [2:0] state; // holds state for sequencing
                            through mults
reg        [7:0] accum; // accumulates multiplier products
reg              clearaccum; // sets accum to zero
reg        [7:0] accumsum;
wire             multdone; // multiplier has completed
wire       [7:0] multout; // multiplier product

// shift-add multiplier for sample-coeff mults
mult8 x 8 mult8 x 8(.clk(clk), .dat1(multdat),
    .dat2(multcoeff), .start(multstart),
    .done(multdone), .multout(multout));
```

example. In this case, we had arbitrary registers that represented the inputs required to create a set of products. The most efficient way to sequence through the set of multiplier inputs was with a state machine.

2.3 RESOURCE SHARING

When we use the term *resource sharing*, we are not referring to the low-level optimizations performed by FPGA place and route tools (this is discussed in later chapters). Instead, we are referring to higher-level architectural resource sharing where different resources are shared across different functional boundaries. This type of resource sharing should be used whenever there are functional blocks that can be used in other areas of the design or even in different modules.

A simple example of resource sharing is with system counters. Many designs use multiple counters for timers, sequencers, state machines, and so forth. Oftentimes, these counters can be pulled to a higher level in the hierarchy and distributed to multiple functional units. For instance, consider modules A and B. Each of these modules uses counters for a different reason. Module A uses the counter to

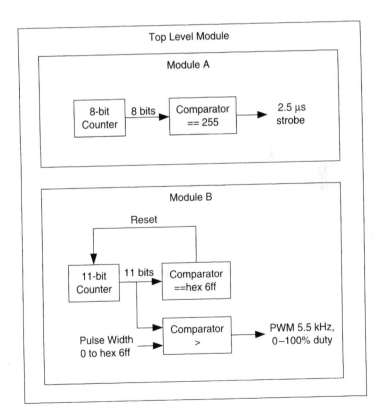

Figure 2.3 Separated counters.

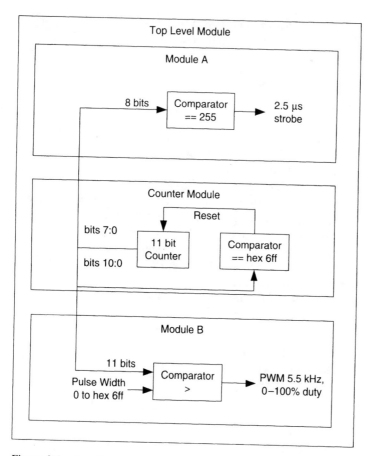

Figure 2.4 Shared counter.

flag an operation every 256 clocks (at 100 MHz, this would correspond with a trigger every 2.56 μs). Module B uses a counter to generate a PWM (Pulse Width Modulated) pulse of varying duty cycle with a fixed frequency of 5.5 kHz (with a 100-MHz system clock, this would correspond with a period of hex 700 clocks).

Each module in Figure 2.3 performs a completely independent operation. The counters in each module also have completely different characteristics. In module A, the counter is 8 bits, free running, and rolls over automatically. In module B, the counter is 11 bits and resets at a predefined value (1666). Nonetheless, these counters can easily be merged into a global timer and used independently by modules A and B as shown in Figure 2.4.

Here we were able to create a global 11-bit counter that satisfied the requirement of both module A and module B.

For compact designs where area is the primary requirement, search for resources that have similar counterparts in other modules that can be brought to a global point in the hierarchy and shared between multiple functional areas.

2.4 IMPACT OF RESET ON AREA

A common misconception is that the reset structures are always implemented in a purely global sense and have little effect on design size. The fact is that there are a number of considerations to take into account relative to area when designing a reset structure and a corresponding number of penalties to pay for a suboptimal design.

The first effect on area has to do with the insistence on defining a global set/reset condition for every flip-flop. Although this may seem like good design practice, it can often lead to a larger and slower design. The reason for this is because certain functions can be optimized according to the fine-grain architecture of the FPGA, but bringing a reset into every synchronous element can cause the synthesis and mapping tools to push the logic into a coarser implementation.

> An improper reset strategy can create an unnecessarily large design and inhibit certain area optimizations.

The next sections describe a number of different scenarios where the reset can play a significant role in the speed/area characteristics and how to optimize accordingly.

2.4.1 Resources Without Reset

This section describes the impact that a global reset will have on FPGA resources that do not have reset available. Consider the following example of a simple shift register:

IMPLEMENTATION 1 : *Synchronous Reset*

```
always @(posedge iClk)
    if(!iReset) sr <= 0;
    else sr        <= {sr[14:0], iDat};
```

IMPLEMENTATION 2 : *No Reset*

```
always @(posedge iClk)
    sr <= {sr[14:0], iDat};
```

The differences between the above two implementations may seem trivial. In one case, the flip-flops have resets defined to be logic-0, whereas in the other implementation, the flip-flops do not have a defined reset state. The key here is that if we wish to take advantage of built-in shift-register resources available in the FPGA, we will need to code it such that there is a direct mapping. If we were targeting a Xilinx device, the synthesis tool would recognize that the shift-register SRL16 could be used to implement the shift register as shown in Figure 2.5.

Note that no resets are defined for the SRL16 device. If resets are defined in our design, then the SRL16 unit could not be used as there are no reset control

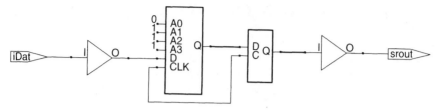

Figure 2.5   Shift register implemented with SRL16 element.

Figure 2.6   Shift register implemented with flip-flops.

Table 2.1   Resource Utilization for Shift Register
Implementations

Implementation	Slices slice	Flip-flops
Resets defined	9	16
No resets defined	1	1

signals to the resource. The shift register would be implemented as discrete flip-flops as shown in Figure 2.6. The difference is drastic as summarized in Table 2.1.

An optimized FPGA resource will not be used if an incompatible reset is assigned to it. The function will be implemented with generic elements and will occupy more area.

By removing the reset signals, we were able to reduce 9 slices and 16 slice flip-flops to a single slice and single slice flip-flop. This corresponds with an optimally compact and high-speed shift-register implementation.

2.4.2   Resources Without Set

Similar to the problem raised in the previous section, some internal resources lack any type of set capability. An example is that of an 8×8 multiplier:

```
module mult8(
    output reg [15:0] oDat,
    input        iReset, iClk,
    input   [7:0] iDat1, iDat2,
    );
```

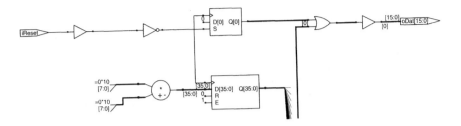

Figure 2.7 Set implemented with external logic.

Table 2.2 Resource Utilization for Set and Reset Implementations

Implementation	Slices slice	Flip-flops	LUTs	Mult16
Reset	9	16	1	1
Set	1	1	1	1

```
always @(posedge iClk)
   if(!iReset) oDat <= 16'hffff;
   else       oDat <= iDat1 * iDat2;

endmodule
```

Again, the only variation to the above code will be the reset condition. Unlike the shift-register example, the multiplier resources in most FPGAs have built-in reset resources. They do not, however, typically have set resources. If the set functionality as described above (16'hffff instead of simply 0) is required, the circuit illustrated in Figure 2.7 will be implemented.

Here an additional gate for each output is required to set the output when the reset is active. The reset on the multiplier, in this case, will go unused. The resource usage between the set and reset implementations is shown in Table 2.2.

By changing the multiplier set to a reset operation, we are able to reduce 9 slices and 16 slice flip-flops to a single slice and single slice flip-flop. This corresponds with an optimally compact and high-speed multiplier implementation.

2.4.3 Resources Without Asynchronous Reset

Many new high-performance FPGAs provide built-in multifunction modules that have general applicability to a wide range of applications. Typically, these resources have some sort of reset functionality but are constrained relative to the type of reset topology. Here we will look at Xilinx-specific multiply–accumulate modules for DSP (Digital Signal Processing) applications. The internal structure of a built-in DSP is typically not flexible to varying reset strategies.

DSPs and other multifunction resources are typically not flexible to varying reset strategies.

Consider the following code for a multiply and accumulate operation:

```
module dspckt(
  output reg [15:0] oDat,
  input      iReset, iClk,
  input      [7:0] iDat1, iDat2);
  reg        [15:0] multfactor;

  always @(posedge iClk or negedge iReset)
    if(!iReset) begin
      multfactor <= 0;
      oDat       <= 0;
    end
    else begin
      multfactor <= (iDat1 * iDat2);
      oDat       <= multfactor + oDat;
    end

endmodule
```

The above code defines a multiply–accumulate function with asynchronous resets. The DSP structures inside a Xilinx Virtex-4 device, for example, have only synchronous reset capabilities as shown in Figure 2.8.

The reset signal here is fed directly into the reset pin of the MAC core. To implement an asynchronous reset as shown in the above code example, on the other hand, the synthesis tool must create additional logic outside of the DSP core.

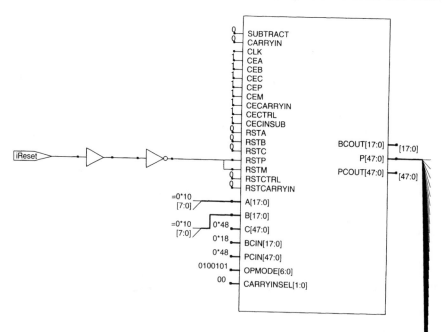

Figure 2.8 Xilinx DSP block with synchronous reset.

Table 2.3 Resource Utilization for Synchronous and
Asynchronous Resets

Architecture	Slices	Flip-flops	LUTs	DSPs
Async Reset	17	32	16	1
Sync Reset	0	0	0	1

Comparing this to a similar structure using synchronous resets, we are able to obtain the results shown in Table 2.3.

When the synchronous reset was used, the synthesis tool was able to use the DSP core available in the FPGA device. By using a different reset than what was available on this device, however, a significant amount of logic was created around it to implement the asynchronous reset.

2.4.4 Resetting RAM

There are reset resources in many built-in RAM (Random Access Memory) resources for FPGAs, but similar to the DSP resource described in the previous sections, often only synchronous resets are available. Attempting to implement an asynchronous reset on a RAM module can be catastrophic to area optimization because there are not smaller elements that can be optimally used to construct a RAM (like a multiplier and an adder can be stitched together to form a MAC module) other than smaller RAM resources, nor can the synthesis tool easily add a few gates to the output to emulate this functionality.

Resetting RAM is usually poor design practice, particularly if the reset is asynchronous.

Consider the following code:

```
module resetckt(
  output reg [15:0] oDat,
  input           iReset, iClk, iWrEn,
  input           [7:0] iAddr, oAddr,
  input           [15:0] iDat);
  reg             [15:0] memdat [0:255];

  always @(posedge iClk or negedge iReset)
  if(!iReset)
    oDat          <= 0;
  else begin
    if(iWrEn)
      memdat[iAddr] <= iDat;

    oDat          <= memdat[oAddr];
  end

endmodule
```

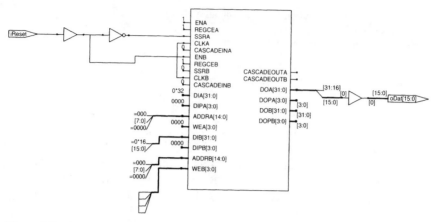

Figure 2.9 Xilinx BRAM with synchronous reset.

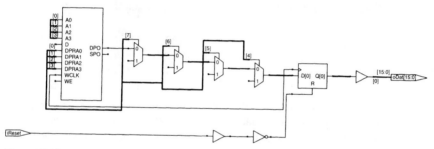

Figure 2.10 Xilinx BRAM with asynchronous reset logic.

Again, the only variation we will consider in the above code is the type of reset: synchronous versus asynchronous. In Xilinx Virtex-4 devices, for example, BRAM (Block RAM) elements have synchronous resets only. Therefore, with a synchronous reset, the synthesis tool will be able to implement this code with a single BRAM element as shown in Figure 2.9.

However, if we attempt to implement the same RAM with an asynchronous reset as shown in the code example above, the synthesis tool will be forced to create a RAM module with smaller distributed RAM blocks, additional decode logic to create the appropriate-size RAM, and additional logic to implement the asynchronous reset as partially shown in Figure 2.10. The final implementation differences are staggering as shown in Table 2.4.

Improperly resetting a RAM can have a catastrophic impact on the area.

Table 2.4 Resource Utilization for BRAM with Synchronous and Asynchronous Resets

Implementation	Slices slice	Flip-flops	4 Input LUTs	BRAMs
Asynchronous reset	3415	4112	2388	0
Synchronous reset	0	0	0	1

2.4.5   Utilizing Set/Reset Flip-Flop Pins

Most FPGA vendors have a variety of flip-flop elements available in any given device, and given a particular logic function, the synthesis tool can often use the set and reset pins to implement aspects of the logic and reduce the burden on the look-up tables. For instance, consider Figure 2.11. In this case, the synthesis tool may choose to implement the logic using the set pin on a flip-flop as shown in Figure 2.12. This eliminates gates and increases the speed of the data path. Likewise, consider a logic function of the form illustrated in Figure 2.13. The AND gate can be eliminated by running the input signal to the reset pin of the flip-flop as shown in Figure 2.14.

The primary reason synthesis tools are prevented from performing this class of optimizations is related to the reset strategy. Any constraints on the reset will not only use available set/reset pins but will also limit the number of library elements to choose from.

Using set and reset can prevent certain combinatorial logic optimizations.

For instance, consider the following implementation in a Xilinx Spartan-3 device:

```
module setreset(
  output reg oDat,
  input     iReset, iClk,
  input     iDat1, iDat2);

  always @(posedge iClk or negedge iReset)
    if(!iReset)
      oDat <= 0;
    else
      oDat <= iDat1 | iDat2;
endmodule
```

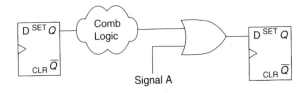

Figure 2.11   Simple synchronous logic with OR gate.

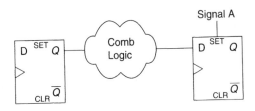

Figure 2.12   OR gate implemented with set pin.

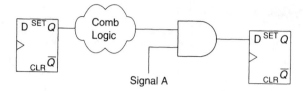

Figure 2.13 Simple synchronous logic with AND gate.

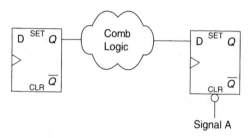

Figure 2.14 AND gate implemented with CLR pin.

In the code example above, an external reset signal is used to reset the state of the flip-flop. This is represented in Figure 2.15.

As can be seen in Figure 2.15, a resetable flip-flop was used for the asynchronous reset capability, and the logic function (OR gate) was implemented in discrete logic. As an alternative, if we remove the reset but implement the same logic function, our design will be optimized as shown in Figure 2.16.

In this implementation, the synthesis tool was able to use the FDS element (flip-flop with a synchronous set and reset) and use the set pin for the OR operation. Thus, by allowing the synthesis tool to choose a flip-flop with a synchronous set, we are able to implement this function with zero logic elements.

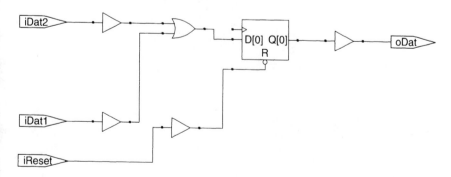

Figure 2.15 Simple asynchronous reset.

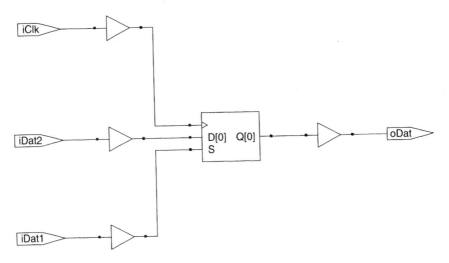

Figure 2.16 Optimization without reset.

We can take this one step further by using both synchronous set and reset signals. If we have a logic equation to evaluate in the form of

$$oDat <= !iDat3 \ \& \ (iDat1 \mid iDat2)$$

we can code this in such a way that both the synchronous set and reset resources are used:

```
module setreset (
  output reg oDat,
  input iClk,
  input iDat1, iDat2, iDat3);

  always @(posedge iClk)
   if(iDat3)
   oDat <= 0;
   else if(iDat1)
    oDat <= 1;
   else
    oDat <= iDat2;

  endmodule
```

Here, the iDat3 input takes priority similar to the reset pin on the associated flip-flops. Thus, this logic function can be implemented as shown in Figure 2.17.

In this circuit, we have three logical operations (invert, AND, and OR) all implemented with a single flip-flop and zero LUTs. Because these optimizations

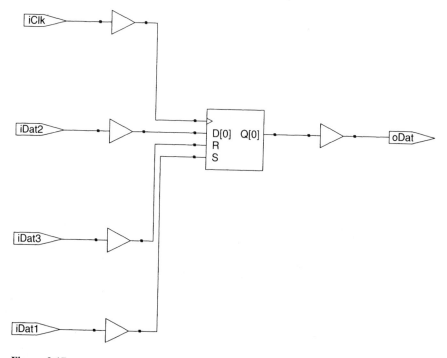

Figure 2.17   Optimization using both set and reset pins.

are not always known at the time the design is architected, avoid using set or reset whenever possible when area is the key consideration.

Avoid using set or reset whenever possible when area is the key consideration.

2.5   SUMMARY OF KEY POINTS

- Rolling up the pipeline can optimize the area of pipelined designs with duplicated logic in the pipeline stages.
- Controls can be used to direct the reuse of logic when the shared logic is larger than the control logic.
- For compact designs where area is the primary requirement, search for resources that have similar counterparts in other modules that can be brought to a global point in the hierarchy and shared between multiple functional areas.
- An improper reset strategy can create an unnecessarily large design and inhibit certain area optimizations.
- An optimized FPGA resource will not be used if an incompatible reset is assigned to it. The function will be implemented with generic elements and will occupy more area.

- DSPs and other multifunction resources are typically not flexible to varying reset strategies.
- Improperly resetting a RAM can have a catastrophic impact on the area.
- Using set and reset can prevent certain combinatorial logic optimizations.
- Avoid using set or reset whenever possible when area is the key consideration.

Chapter 3

Architecting Power

This chapter discusses the third of three primary physical characteristics of a digital design: power. Here we also discuss methods for architectural power optimization in an FPGA.

Relative to ASICs (application specific integrated circuits) with comparable functionality, FPGAs are power-hungry beasts and are typically not well suited for ultralow-power design techniques. A number of FPGA vendors do offer low-power CPLDs (complex programmable logic devices), but these are very limited in size and capability and thus will not always fit an application that requires any respectable amount of computing power. This section will discuss techniques to maximize the power efficiency of both low-power CPLDs as well as general FPGA design.

In CMOS technology, dynamic power consumption is related to charging and discharging parasitic capacitances on gates and metal traces. The general equation for current dissipation in a capacitor is

$$I = V * C * f$$

where I is total current, V is voltage, C is capacitance, and f is frequency.

Thus, to reduce the current drawn, we must reduce one of the three key parameters. In FPGA design, the voltage is usually fixed. This leaves the parameters C and f to manipulate the current. The capacitance C is directly related to the number of gates that are toggling at any given time and the lengths of the routes connecting the gates. The frequency f is directly related to the clock frequency. All of the power-reduction techniques ultimately aim at reducing one of these two components.

During the course of this chapter, we will discuss the following topics:

- The impact of clock control on dynamic power consumption
- Problems with clock gating
    Managing clock skew on gated clocks
- Input control for power minimization

Advanced FPGA Design. By Steve Kilts
Copyright © 2007 John Wiley & Sons, Inc.

- Impact of the core voltage supply
- Guidelines for dual-edge triggered flip-flops
- Reducing static power dissipation in terminations

Reducing dynamic power dissipation by minimizing the route lengths of high toggle rate nets requires a background discussion of placement and routing, and is therefore discussed in Chapter 15 Floorplanning.

3.1 CLOCK CONTROL

The most effective and widely used technique for lowering the dynamic power dissipation in synchronous digital circuits is to dynamically disable the clock in specific regions that do not need to be active at particular stages in the data flow. Since most of the dynamic power consumption in an FPGA is directly related to the toggling of the system clock, temporarily stopping the clock in inactive regions of the design is the most straightforward method of minimizing this type of power consumption. The recommended way to accomplish this is to use either the clock enable pin on the flip-flop or to use a global clock mux (in Xilinx devices this is the BUFGMUX element). If these clock control elements are not available in a particular technology, designers will sometimes resort to direct gating of the system clock. Note that this is not recommended for FPGA designs, and this section describes the issues involved with direct gating of the system clock.

> Clock control resources such as the clock enable flip-flop input or a global clock mux should be used in place of direct clock gating.

Note that this section assumes the reader is already familiar with general FPGA clocking guidelines. In general, FPGAs are synchronous devices, and a number of difficulties arise when multiple domains are introduced through gating or asynchronous interfaces. For a more in-depth discussion regarding clock domains, see Chapter 6.

Figure 3.1 illustrates the poor design practice of simple clock gating. With this clock topology, all flip-flops and corresponding combinatorial logic is active (toggling) whenever the Main Clock is active. The logic within the dotted box, however, is only active when Clock Enable = 1. Here, we refer to the Clock Enable signal as the gating or enable signal. By gating portions of circuitry as shown above, the designer is attempting to reduce the dynamic power dissipation proportional to the amount of logic (capacitance C) and the average toggle frequency of the corresponding gates (frequency f).

> Clock gating is a direct means for reducing dynamic power dissipation but creates difficulties in implementation and timing analysis.

Before we proceed to the implementation details, it is important to note how important careful clock planning is in FPGA design. The system clock is central

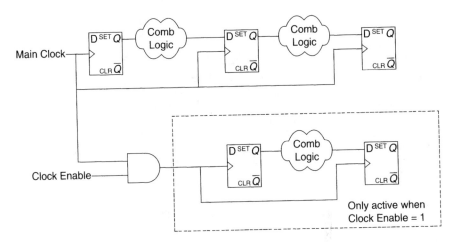

Figure 3.1   Simple clock gating: poor design practice.

to all synchronous digital circuits. EDA (electronic design automation) tools are driven by the system clock to optimize and validate synthesis, layout, static timing analysis, and so forth. Thus, the system clock or clocks are sacred and must be characterized up front to drive the implementation process. Clocks are even more sacred in FPGAs than they are in ASICs, and thus there is less flexibility relative to creative clock structures.

When a clock is gated even in the most trivial sense, the new net that drives the clock pins is considered a new clock domain. This new clock net will require a low-skew path to all flip-flops in its domain, similar to the system clock from which it was derived. For the ASIC designer, these low-skew lines can be built in the custom clock tree, but for the FPGA designer this presents a problem due to the limited number and fixed layout of the low-skew lines.

A gated clock introduces a new clock domain and will create difficulties for the FPGA designer.

The following sections address the issues introduced by gated clocks.

3.1.1   Clock Skew

Before directly addressing the issues related to gated clocks, we must first briefly review the topic of clock skew. The concept of clock skew is a very important one in sequential logic design.

In Figure 3.2, the propagation delay of the clock signal between the first flip-flop and the second flip-flop is assumed to be zero. If there is positive delay through the cloud of combinatorial logic, then timing compliance will be determined by the clock period relative to the combinatorial delay + logic routing

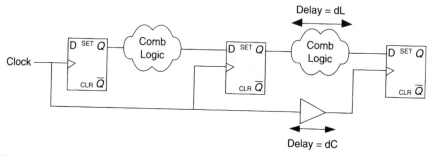

Figure 3.2   Clock skew.

delay + flip-flop setup time. A signal can only propagate between a single set of flip-flops for every clock edge. The situation between the second and third flip-flop stages, however, is different. Because of the delay on the clock line between the second and third flip-flops, the active clock edge will not occur simultaneously at both elements. Instead, the active clock edge on the third flip-flop will be delayed by an amount dC.

If the delay through the logic (defined as dL) is less than the delay on the clock line (dC), then a situation may occur where a signal that is propagated through the second flip-flop will arrive at the third stage before the active edge of the clock. When the active edge of the clock arrives, the same signal could be propagated through stage 3. Thus, a signal could propagate through both stage 2 and stage 3 on the same clock edge! This scenario will cause a catastrophic failure of the circuit, and thus clock skew must be taken into account when performing timing analysis. It is also important to note that clock skew is independent of clock speed. The "fly-through" issue described above will occur exactly the same way regardless of the clock frequency.

Mishandling clock skew can cause catastrophic failures in the FPGA.

3.1.2 Managing Skew

Low-skew resources provided on FPGAs ensure that the clock signal will be matched on all clock inputs as tightly as possible (within picoseconds). Take, for instance, the scenario where a gate is introduced to the clock network as shown in Figure 3.3.

The clock line must be removed from the low-skew global resource and routed to the gating logic, in this case an AND gate. The fundamental problem of adding skew to the clock line is now the same as it was in the problem described previously. It is conceivable that the delay through the gate (dG) plus the routing delays will be greater than the delay through the logic (dL). To handle this potential problem, the implementation and analysis tools must be given a set of constraints such that any timing problems associated with skew through the gating item are eliminated and then analyzed properly in post-implementation analysis.

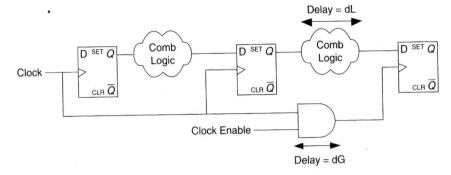

Figure 3.3 Clock skew introduced with clock gating: Poor design practice.

As an example, consider the following module that uses clock gating:

```
// Poor design practice
module clockgating(
   output dataout,
   input   clk, datain,
   input   clockgate1);
   reg     ff0, ff1, ff2;
   wire    clk1;

   // clocks are disabled when gate is low
   assign clk1    = clk & clockgate1;
   assign dataout = ff2;

   always @(posedge clk)
      ff0 <= datain;

   always @(posedge clk)
      ff1 <= ff0;

   always @(posedge clk1)
      ff2 <= ff1;
endmodule
```

In the above example, there is no logic between the flip-flops on the data path, but there is logic in the clock path as shown in Figure 3.4.

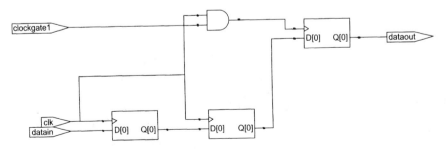

Figure 3.4 Clock skew as the dominant delay.

Different tools handle this situation differently. Some tools such as Synplify will remove the clock gating by default to create a purely synchronous design. Other tools ignore skew problems if the clocks remain unconstrained but will add artificial delays once the clocks have been constrained properly.

Unlike ASIC designs, hold violations in FPGA designs are rare due to the built-in delays of the logic blocks and routing resources. One thing that can cause a hold delay, however, is excessive delay on the clock line as shown above. Due to the fact that the data propagates in less than 1 ns and the clock in almost 2 ns, the data will arrive almost 1 ns before the clock and lead to a serious timing violation. Depending on the synthesis tool, this can sometimes be fixed by adding a clock constraint. A subsequent analysis may or may not show (depending on the technology) that artificial routing delay was added to the data path to eliminate the hold violation.

Clock gating can cause hold violations that may or may not be corrected by the implementation tools.

It is again worth reiterating that most vendors have advanced clock buffer technology that provide enable capability to certain branches of the clock tree. This type of control is always recommended above clock gating with logic elements.

3.2   INPUT CONTROL

An often overlooked power-reduction technique is that of input slew rates. CMOS input buffers can create excessive current draw under conditions where both the high-side and low-side transistors are conducting at the same time. To conceptualize this, consider a basic first-order model of a CMOS transistor that describes I_{ds} in terms of V_{ds} as illustrated in Figure 3.5, where the regions are defined by:

Cutoff: $V_{gs} < V_{th}$

Linear (resistive): $0 < V_{ds} < V_{gs} - V_{th}$

Saturation: $0 < V_{gs} - V_{th} < V_{ds}$

where V_{gs} is the gate-to-source voltage, V_{th} is the device threshold voltage, and V_{ds} is the drain-to-source voltage.

An ideal switching scheme would be one where the input to a gate switched from cutoff to the linear region instantaneously, and the complementary logic switched the opposite direction at the same instant. If one of the two complements is always in cutoff, there is no current flowing through both sides of the logic gate at the same time (and thus providing a resistive path between power and ground). For an inverter, this would mean that the NMOS (N-channel MOSFET) device would transition from 0 to VDD (positive power rail) taking the NMOS from cutoff to the linear region instantly, and the PMOS (P-channel MOSFET) would transition from the linear region to cutoff at the same instant. In the opposite transition when V_{gs} transitions from VDD to 0, the NMOS would move from the linear region to cutoff instantly, and the PMOS would move from the cutoff region to the linear region at the same instant.

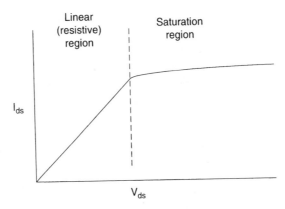

Figure 3.5   Simple I/V curve for a CMOS transistor.

In a real system, however, we must take into consideration the transition times and the behavior of the transistors during those transitions. For instance, consider a CMOS inverter that has an input of 0 V and an output of VDD. As the input transitions from 0 to VDD (a 0 to 1 transition), the NMOS transistor leaves the cutoff region as soon as the input passes the threshold V_{th} and enters into the saturation region. The PMOS device is still in the linear region during the early part of this transition, and so current begins to flow between VDD and ground. As the input rises, the output falls. When the drain of the NMOS falls below a threshold of the gate voltage, the NMOS transitions into the linear region, and the PMOS transitions to saturation and then to cutoff. To minimize the power dissipation, it is desirable to minimize the time in the saturation region; that is, minimize the time during which the gate inputs are transitioning.

> To minimize the power dissipation of input devices, minimize the rise and fall times of the signals that drive the input.

Another important conclusion can be drawn from the above equations. If the driving signal is not within a threshold voltage of 0 or V_{dd} in steady state (i.e., when the gate is not switching), the transistor previously in cutoff will enter into the saturation region and begin to dissipate a small amount of current. This can be a problem in systems where smaller signal swings are used to drive inputs that are powered by a higher voltage.

In harmony with the principle described above, a floating input may be an even worse problem than an underdriven input. A floating input is by definition an underdriven input, but because it is floating there is no way to know how underdriven it is. It may be that the input has settled at a metastable point where both transistors are in the saturation region. This would have disastrous implications relative to power dissipation. Worse yet, this would not be a repeatable problem. Because most FPGA devices have resistive terminations available for unused inputs, it is good design practice to define a logic state for these and avoid the unpredictable effects of floating inputs.

> Always terminate unused input buffers. Never let an FPGA input buffer float.

3.3 REDUCING THE VOLTAGE SUPPLY

Although reducing the supply voltage is usually not a desirable option, it is worth mentioning due to the dramatic effect it can have on power consumption. Power dissipation in a simple resistor will drop off with the square of the voltage. Thus, significant power savings can be achieved by lowering the power supply voltage of the FPGA near the minimum required voltage. It is important to note, however, that lowering the voltage will also decrease the performance of the system. If this method is used, ensure that the timing analysis takes into consideration the lowest possible voltage on the supply rail for worst-case maximum timing.

> Dynamic power dissipation drops off with the square of the core voltage, but reducing voltage will have a negative impact on performance.

Because the core voltage on an FPGA will be rated from 5% to 10% of the specified value, great care must be given to this from a system perspective. Typically, power issues can be addressed with other techniques while keeping the core voltage well within the specified range.

3.4 DUAL-EDGE TRIGGERED FLIP-FLOPS

Due to the fact that power dissipation is proportional to the frequency that a signal toggles, it is desirable to maximize the amount of functionality for each toggle of a high fan-out net. Most likely, the highest fan-out net is the system clock, and thus any techniques to reduce the frequency of this clock would have a dramatic impact on dynamic power consumption. Dual-edge triggered flip-flops provide a mechanism to propagate data on both edges of the clock instead of just one. This allows the designer to run a clock at half the frequency that would otherwise be required to achieve a certain level of functionality and performance.

Coding a dual-edge triggered flip-flop is very straightforward. The following example illustrates this with a simple shift register. Note that the input signal is captured on the rising edge of the clock and is then passed to dual-edge flip-flops.

```
module dualedge(
   output reg dataout,
   input      clk, datain);
   reg        ff0, ff1;

   always @(posedge clk)
      ff0     <= datain;

   always @(posedge clk or negedge clk) begin
      ff1     <= ff0;
      dataout <= ff1;
      end
endmodule
```

Note that if dual-edge flip-flops are not available, redundant flip-flops and gating will be added to emulate the appropriate functionality. This could completely defeat the purpose of using the dual-edge strategy and should be analyzed appropriately after implementation. A good synthesis tool will at least flag a warning if no dual-edge devices are available.

> Dual-edge triggered flip-flops should only be used if they are provided as primitive elements.

The Xilinx Coolrunner-II family includes a feature named CoolClock, which divides the incoming clock by 2 and then switches the flip-flops to dual-edge devices as described above. From an external perspective, the device behaves the same as a single-edge triggered system but with half of the dynamic power dissipation on the global clock lines.

3.5 MODIFYING TERMINATIONS

Resistive loads connected to output pins are common in systems with bus signals, open-drain outputs, or transmission lines requiring termination. In all of these cases, one of the CMOS transistors on the output driver of the FPGA will need to source or sink current through one of these resistive loads. For outputs requiring pull-up resistors, calculate the minimum acceptable rise-time to size the resistor as large as possible. If there are high side drivers as well as low side drivers, ensure there is never a condition where bus contention occurs as this will draw excessive currents even if for only a few nanoseconds at a time. For transmission lines with shunt termination at the load, a series termination may be used as an alternate depending on the requirements of the system. As can be seen in Figure 3.6, there is not steady-state current dissipation with a series termination.

> There is no steady-state current dissipation with a series termination.

The disadvantages are

- An initial reflection from the load to the terminating resistor
- A small amount of attenuation through the series resistor during a transition

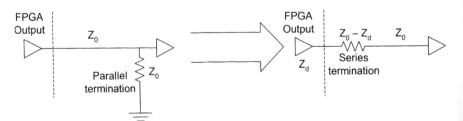

Figure 3.6   Termination types.

If these performance characteristics are acceptable for a given system, the series termination approach will eliminate static power dissipation through the termination resistor.

3.6 SUMMARY OF KEY POINTS

- Clock control resources such as the clock enable flip-flop input or a global clock mux should be used in place direct clock gating when they are available.
- Clock gating is a direct means for reducing dynamic power dissipation but creates difficulties in implementation and timing analysis.
- Mishandling clock skew can cause catastrophic failures in the FPGA.
- Clock gating can cause hold violations that may or may not be corrected by the implementation tools.
- To minimize the power dissipation of input devices, minimize the rise and fall times of the signals that drive the input.
- Always terminate unused input buffers. Never let an FPGA input buffer float.
- Dynamic power dissipation drops off with the square of the core voltage, but reducing voltage will have a negative impact on performance.
- Dual-edge triggered flip-flops should only be used if they are provided as primitive elements.
- There is no steady-state current dissipation with a series termination.

Chapter 4

Example Design: The Advanced Encryption Standard

The Advanced Encryption Standard (AES; also referred to by its original name, Rijndael) specifies the latest standard in encryption for the protection of electronic information. The standard has been approved by the U.S. National Institute of Standards and Technology (NIST), which has made the specification publicly available in a Federal Information Processing Standards Publication (FIPS PUB 197). The motivation behind the new standard was the weakness of the existing Data Encryption Standard (DES). In addition to providing more security, AES is designed to lend itself to an easy implementation in hardware. In this context, easier means less prone to design error (more reliable) and faster (simple combinatorial logic).

The objective of this chapter is to describe a number of AES architectures and to analyze the various trade-offs relative to performance versus area.

4.1 AES ARCHITECTURES

AES is a symmetric, secret-key cipher that maps a 128-bit block of plaintext data to a 128-bit block of ciphertext. The length of the key is variable between 128, 192, and 256 bits and will determine level of security (longer key = larger key space = more security). The transformations in the AES algorithm consist of four components organized as distinct modules: Sub Bytes (bit mapping), shift rows (swapping), mult-column [transformation over $GF(2^8)$], and Add Round Key [addition of round key with bitwise operations in the field GF(2)]. These transformations make up a "round," and the number of rounds is determined by the key size (128 bits, 10 rounds; 192 bits, 12 rounds; 256 bits, 14 rounds). Note that

Advanced FPGA Design. By Steve Kilts
Copyright © 2007 John Wiley & Sons, Inc.

the round key for each round is unique. These round keys are derived from the original key through the key expansion. The key expansion is one of the architectural focal points of this chapter and will be discussed in more detail. For a more detailed explanation of the complete AES cipher, see the Federal Information Processing Standard 197 (FIPS 197), as provided by NIST.

The key expansion, which runs parallel to the data path, takes the cipher key and creates a unique key for each transformation round. Let a word $= 32$ bits and $N_k = $ Keysize/Wordsize ($=128$, 192, or $256/32$). The first N_k words of the expanded key are filled with the cipher key. Every subsequent 32-bit word in the expanded key is the XOR (Exclusive-OR) of the previous 32-bit word and the 32-bit-word N_k words previous to the current word. For words that occur on a multiple of N_k, the current word undergoes a transformation prior to the XOR operation, followed by an XOR with a round constant. The transformation consists of a cyclic permutation, followed by an 8-byte mapping for all four bytes in the 32-bit word. The round constant is defined by FIPS 197 as the values given by $[x^{(i-1)}, \{00\}, \{00\}, \{00\}]$, with $x^{(i-1)}$ being powers of x, where x is denoted as $\{02\}$ in the field $GF(2^8)$.

A single key expansion operation is autonomous relative to the high-level architecture and is shown in the following implementation.

```
module KeyExp1Enc(
   // updated values to be passed to next iteration
   output [3:0]            oKeyIter, oKeyIterModNk,
                          oKeyIterDivNk,
   output [32*'Nk-1:0]     oNkKeys,
   input                  iClk, iReset,
   // represents total # of iterations and value mod Nk
   input [3:0]             iKeyIter, iKeyIterModNk,
                          iKeyIterDivNk,
   // The last Nk keys generated in key expansion
   input [32*'Nk-1:0]     iNkKeys);
   // updated values to be passed to next iteration
   reg [3:0]               oKeyIter, oKeyIterModNk,
                          oKeyIterDivNk;
   reg [32*'Nk-1:0]        OldKeys;
   reg [31:0]              InterKey; // intermediate key value
   wire [32*'Nk-1:0]       oNkKeys;
   wire [31:0]             PrevKey, RotWord, SubWord,
                          NewKeyWord;
   wire [31:0]             KeyWordNk;
   wire [31:0]             Rcon;

   assign PrevKey      =    iNkKeys[31:0]; // last word in key
                          array
   assign KeyWordNk    =    OldKeys[32*'Nk-1:32*'Nk-32];

   // 1 byte cyclic permutation
   assign RotWord      =    {PrevKey[23:0], PrevKey[31:24]};
```

```
// new key calculated in this round
assign NewKeyWord  = KeyWordNk ^ InterKey;

// calculate new key set
assign oNkKeys     = {OldKeys[32*'Nk-33:0], NewKeyWord};

// calculate Rcon over GF(2^8)
assign Rcon        = iKeyIterDivNk == 8'h1 ? 32'h01000000:
                     iKeyIterDivNk == 8'h2 ? 32'h02000000:
                     iKeyIterDivNk == 8'h3 ? 32'h04000000:
                     iKeyIterDivNk == 8'h4 ? 32'h08000000:
                     iKeyIterDivNk == 8'h5 ? 32'h10000000:
                     iKeyIterDivNk == 8'h6 ? 32'h20000000:
                     iKeyIterDivNk == 8'h7 ? 32'h40000000:
                     iKeyIterDivNk == 8'h8 ? 32'h80000000:
                     iKeyIterDivNk == 8'h9 ? 32'h1b000000:
                     32'h36000000;
SboxEnc SboxEnc0(.iPreMap(RotWord[31:24]),
.oPostMap(SubWord[31:24]));
SboxEnc SboxEnc1(.iPreMap(RotWord[23:16]),
.oPostMap(SubWord[23:16]));
SboxEnc SboxEnc2(.iPreMap(RotWord[15:8]),
.oPostMap(SubWord[15:8]));
SboxEnc SboxEnc3(.iPreMap(RotWord[7:0]),
.oPostMap(SubWord[7:0]));

'ifdef Nk8

wire [31:0] SubWordNk8;

// Substitution only when Nk = 8
SboxEnc SboxEncNk8_0(.iPreMap(PrevKey[31:24]),
.oPostMap(SubWordNk8[31:24]));
SboxEnc SboxEncNk8_1(.iPreMap(PrevKey[23:16]),
.oPostMap(SubWordNk8[23:16]));
SboxEnc SboxEncNk8_2(.iPreMap(PrevKey[15:8]),
.oPostMap(SubWordNk8[15:8]));
SboxEnc SboxEncNk8_3(.iPreMap(PrevKey[7:0]),
.oPostMap(SubWordNk8[7:0]));

'endif

always @(posedge iClk)
  if(!iReset) begin
    oKeyIter          <= 0;
    oKeyIterModNk     <= 0;
    InterKey          <= 0;
    oKeyIterDivNk     <= 0;
    OldKeys           <= 0;
  end
  else begin
```

```
oKeyIter                        <= iKeyIter + 1;
OldKeys                         <= iNkKeys;

// update "Key iteration mod Nk" for next iteration
if(iKeyIterModNk + 1 == 'Nk) begin
   oKeyIterModNk                <= 0;
   oKeyIterDivNk                <= iKeyIterDivNk+1;
end
else begin
   oKeyIterModNk                <= iKeyIterModNk + 1;
   oKeyIterDivNk                <= iKeyIterDivNk;
end

if(iKeyIterModNk == 0)
   InterKey                     <= SubWord ^ Rcon;
'ifdef Nk8
// an option only for Nk = 8
else if(iKeyIterModNk == 4)
   InterKey                     <= SubWordNk8;
'endif
else
   InterKey                     <= PrevKey;
end
endmodule
```

Likewise, the autonomous operation for the data path is the combination of all functions required for a round encryption as shown in the following implementation.

```
module RoundEnc(
   output [32*'Nb-1:0]   oBlockOut,
   output                oValid,
   input                 iClk, iReset,
   input  [32*'Nb-1:0]   iBlockIn, iRoundKey,
   input                 iReady,
   input  [3:0]          iRound);
   wire   [32*'Nb-1:0]   wSubOut, wShiftOut, wMixOut;
   wire                  wValidSub, wValidShift,
                         wValidMix;

SubBytesEnc sub(         .iClk(iClk), .iReset(iReset),
                         .iBlockIn(iBlockIn),
                         .oBlockOut(wSubOut),
                         .iReady(iReady),
                         .oValid(wValidSub));

ShiftRowsEnc shift(      .iBlockIn(wSubOut), .oBlock
                            Out(wShiftOut),
                         .iReady(wValidSub), .oValid
                            (wValidShift));

MixColumnsEnc mixcolumn( .iClk(iClk), .iReset(iReset),
                         .iBlockIn(wShiftOut),
```

```
                             .oBlockOut(wMixOut),
                             .iReady(wValidShift),
                              .oValid(wValidMix),
                              .iRound(iRound));

AddRoundKeyEnc addroundkey(.iClk(iClk),  .iReset(iReset),
                             .iBlockIn(wMixOut),
                             .iRoundKey(iRoundKey),
                             .oBlockOut(oBlockOut),
                             .iReady(wValidMix),
                              .oValid(oValid));
endmodule
```

The implementation of the Round subblocks is straightforward. For reasons described later, assume that each Round has a latency of 4 clocks. It would be reasonable (based on logic balancing) to distribute the pipeline stages as follows.

4.1.1 One Stage for Sub-bytes

Sub-bytes is implemented as a look-up table due to the iterative nature of the algorithms implemented by it as well as the relatively small map space. Thus, an 8-bit to 8-bit mapping would be efficiently implemented as a synchronous 8×256 (2^8) ROM with a single pipeline stage. This is shown in Figure 4.1.

4.1.2 Zero Stages for Shift Rows

This stage simply mixes the rows in the data block, so no logic is used here. Thus, another pipeline stage at this point would create an imbalance of logic

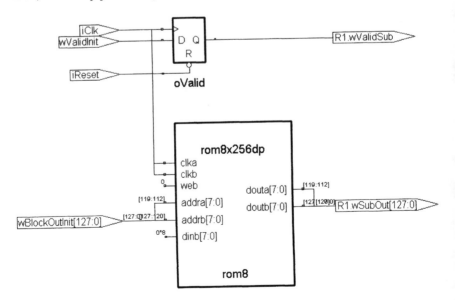

Figure 4.1 An 8-bit mapping in the sub-bytes module.

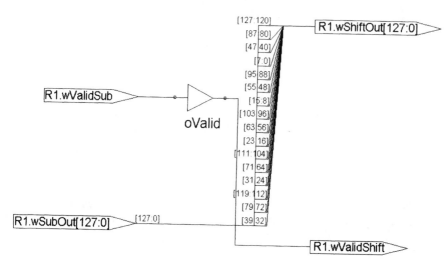

Figure 4.2   Shift-row implementation.

around the pipeline stages and thus decrease maximum frequency and total throughput. The shifted rows are illustrated in Figure 4.2.

4.1.3   Two Pipeline Stages for Mix-Column

This stage has the most logic out of all four Round stages and is thus the best place to add the additional pipeline stage. The mix-column hierarchy is shown in Figure 4.3.

As can be seen from Figure 4.3, Mix-Column uses a module called Map-Column as a building block. This can be seen in Figure 4.4.

As can be seen from Figure 4.4, Map-Column uses a block called Poly-Mult X2 (polynomial $\times 2$ multiplier) as a building block. This is shown in Figure 4.5.

4.1.4   One Stage for Add Round Key

This stage simply XORs the round key from the key expansion pipeline with the data block. This is shown in Figure 4.6.

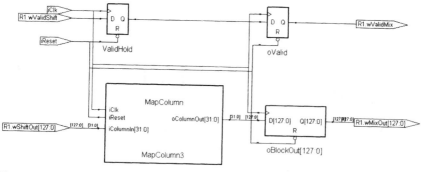

Figure 4.3   The mix-column hierarchy.

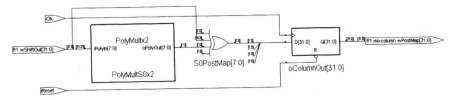

Figure 4.4 The Map-column hierarchy.

4.1.5 Compact Architecture

The first implementation under consideration is a compact implementation designed to iteratively reuse logic resources. Initially, the incoming data and key are added together in the Initial Round module, and the result is registered before entering the encryption loop. The data is then applied to the Sub Bytes, Shift Rows, Mult-Column, and Add Round Key in the specified order. At the end of each round, the new data is registered. These operations are repeated according to the number of rounds. A block diagram of the iterative architecture is shown in Figure 4.7.

The top-level implementation is shown in the following code.

```
module AES_Enc_core(
    output [32*'Nb-1:0]    oCiphertext, // output cipthertext
    output                 oValid, // data at output is valid
    output                 oKeysValid,
```

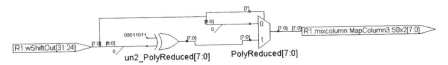

Figure 4.5 Polynomial multiplication x2.

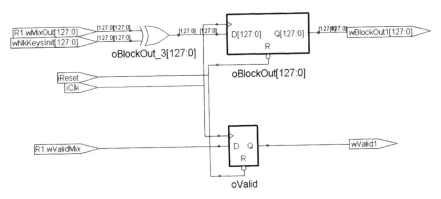

Figure 4.6 The add-round-key block.

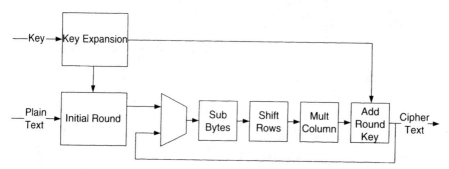

Figure 4.7   A compact implementation.

```
input                    iClk, iReset,
input [32*'Nb-1:0]       iPlaintext, // input data to be
   encrypted
input [32*'Nk-1:0]       iKey, // input cipther key
input                    iReady, // valid data to encrypt
input                    iNewKey); // signals new key is
input
// registered inputs
wire    [32*'Nk-1:0]     wKeyReg;
wire                     wNewKeyReg, wReadyReg;
wire    [127:0]            wPlaintextReg, wBlockOutInit;
wire    [127:0]            wRoundKeyInit, wRoundKey;

// register inputs
InputRegsEnc InputRegs( .iClk(iClk), .iReset(iReset),
                        .iKey(iKey),
                        .iNewKey(iNewKey), .iPlaintext
                        (iPlaintext),
                        .oKeysValid(oKeysValid),
                          .iReady(iReady),
                        .oKey(wKeyReg), .oPlaintext
                        (wPlaintextReg),
                        .oReady(wReadyReg));

// initial addition of round key
AddRoundKeyEnc InitialKey( .iClk(iClk), .iReset(iReset),
                          .iBlockIn(wPlaintextReg),
                          .iRoundKey(wRoundKeyInit),
                          .oBlockOut(wBlockOutInit),
                          .iReady(wReadyReg),
                          .oValid(wValidInit));

// Number of rounds is a function of key size (10, 12, or
   14)
// Key expansion block
```

```
KeyExpansionEnc KeyExpansion( .iClk(iClk), .iReset
                            (iReset),
                 .iNkKeys(wKeyReg),
                  .iReady(wReadyReg),
                 .oRoundKey(wRoundKey));

RoundsIterEnc RoundsIter(    .iClk(iClk), .iReset(iReset),
                .iBlockIn(wBlockOutInit),
                .oBlockOut(oCiphertext),
                .iReady(wValidInit),
                 .oValid(oValid),
                .iRoundKey(wRoundKey));

`ifdef Nk4
assign wRoundKeyInit = wKeyReg[128-1:0];
`endif

`ifdef Nk6
assign wRoundKeyInit = wKeyReg[192-1:192-128];
`endif

`ifdef Nk8
assign wRoundKeyInit = wKeyReg[256-1:256-128];
`endif
endmodule
```

In the above code, the modules KeyExpansionEnc and RoundsIterEnc perform the iterative operations required of the compact architecture. The KeyExpansionEnc handles the iterations for the key expansion, and RoundsIterEnc handles the data path. A unique round key is passed from the key expansion module to RoundsIterEnc for every iteration of the round. The following code loops the key information through the same expansion module to reuse the logic for every round:

```
module KeyExpansionEnc(
    output [128-1:0]        oRoundKey,
    input                   iClk, iReset,
    // The last Nk keys generated in initial key expansion
    input  [32*`Nk-1:0]     iNkKeys,
    input                   iReady); // signals a
                                     new key is input

    wire   [3:0]            KeyIterIn, KeyIterOut;
    wire   [3:0]            KeyIterDivNkIn, KeyIterDivNkOut;
    wire   [3:0]            KeyIterModNkIn, KeyIterModNkOut;
    wire   [32*`Nk-1:0]     NkKeysOut, NkKeysIn;
    wire                    wReady;

    assign wReady      =    iReady;
    assign KeyIterIn   =    wReady ? `Nk : KeyIterOut;
    assign oRoundKey   =    NkKeysOut[32*`Nk-1:32*`Nk-128];
```

```
assign KeyIterModNkIn =  wReady ? 4'h0    : KeyIter
                               ModNkOut;
assign KeyIterDivNkIn =  wReady ? 4'h1    : KeyIter
                               DivNkOut;
assign NkKeysIn       =  wReady ? iNkKeys : NkKeysOut;

KeyExp1Enc KeyExp1(.iClk(iClk), .iReset(iReset),
             .iKeyIter(KeyIterIn),
             .iKeyIterModNk(KeyIterModNkIn),
             .iNkKeys(NkKeysIn), .iKeyIterDivNk
               (KeyIterDivNkIn),
             .oKeyIter(KeyIterOut),
             .oKeyIterModNk(KeyIterModNkOut),
             .oNkKeys(NkKeysOut),
             .oKeyIterDivNk(KeyIterDivNkOut));
endmodule
```

In the above module, the output of the single key expansion module KeyExp1Enc is routed back to the input for further expansion on subsequent rounds as shown in Figure 4.8.

Thus, the logic in the KeyExp1Enc is reused for every round.

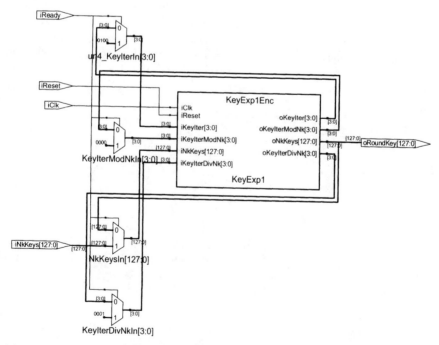

Figure 4.8 Iterative key expansion.

4.1.6 Partially Pipelined Architecture

The second implementation under consideration is a partially pipelined architecture. An AES round is completed in 11 to 14 clock cycles depending on the key size.

As can be seen from Figure 4.9, multiple instantiations of the data path core can be used to create a pipelined design, where the key expansion is performed in a static fashion. This implementation is shown in the following code for $N_k = 4$.

```
module AES_core(
    output [32*'Nb-1:0]    oCiphertext, // output cipthertext
    output                 oValid, // data at output is valid
    // signals that new key has been completely processed
    output                 oKeysValid,
    input                  iClk, iReset,
    input [32*'Nb-1:0]     iPlaintext, // input data to
                           be encrypted
    input [32*'Nk-1:0]     iKey, // input cipther key
    input                  iReady, // valid data to encrypt
    input                  iNewKey); // signals new key
                           is input
    wire [32*'Nb-1:0]      wRoundKey1, wRoundKey2,
                           wRoundKey3, wRoundKey4,
                           wRoundKey5, wRoundKey6,
                           wRoundKey7, wRoundKey8,
                           wRoundKey9, wRoundKeyFinal,
                           wRoundKeyInit;
    wire [32*'Nb-1:0]      wBlockOut1, wBlockOut2,
                           wBlockOut3, wBlockOut4,
                           wBlockOut5, wBlockOut6,
```

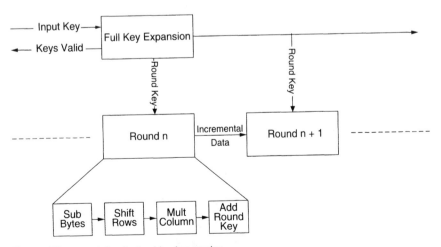

Figure 4.9 A partially pipelined implementation.

```
                                   wBlockOut7, wBlockOut8,
                                   wBlockOut9, wBlockOutInit;
wire  [32*'Nk-1:0]                 wNkKeysInit;
wire  [3:0]                        wKeyIterInit;
wire  [3:0]                        wKeyIterModNkInit;
wire  [3:0]                        wKeyIterDivNkInit;
wire                               wValid1, wValid2, wValid3,
                                     wValid4,
                                   wValid5, wValid6, wValid7,
                                     wValid8,
                                   wValid9, wValidFinal,
                                     wValidInit;
wire                               wNewKeyInit;
wire  [128*('Nr+1)-1:0]            wKeys; // complete set
                                   of round keys

// registered inputs
wire  [32*'Nk-1:0]                 wKeyReg;
wire                               wNewKeyReg, wReadyReg;
wire  [127:0]                      wPlaintextReg;

// register inputs
InputRegs InputRegs(               .iClk(iClk), .iReset(iReset),
                                     .iKey(iKey),
                                   .iNewKey(iNewKey),
                                   .iPlaintext(iPlaintext),
                                   .iReady(iReady), .oKey(wKeyReg),
                                   .oNewKey(wNewKeyReg),
                                   .oPlaintext(wPlaintextReg),
                                   .oReady(wReadyReg));

// initial key expansion
KeyExpInit KeyExpInit(             .iClk(iClk), .iReset(iReset),
                                   .iNkKeys(wKeyReg), .iNewKey
                                     (wNewKeyReg),
                                   .oKeyIter(wKeyIterInit),
                                   .oNewKey(wNewKeyInit),
                                   .oKeyIterModNk
                                     (wKeyIterModNkInit),
                                   .oNkKeys(wNkKeysInit),
                                   .oKeyIterDivNk
                                     (wKeyIterDivNkInit));

// initial addition of round key
AddRoundKey InitialKey(  .iClk(iClk), .iReset(iReset),
                                   .iBlockIn(wPlaintextReg),
                                   .iRoundKey(wRoundKeyInit),
                                   .oBlockOut(wBlockOutInit),
                                       .iReady(wReadyReg),
                                       .oValid(wValidInit));
```

```
// Number of rounds is a function of key size (10, 12, or
   14)

// Key expansion block
KeyExpansion KeyExpansion(   .iClk(iClk),
                             .iReset(iReset),
                             .iKeyIter(wKeyIterInit),
                             .iKeyIterModNk(wKeyIterMod
                                NkInit),
                             .iNkKeys(wNkKeysInit),
                             .iKeyIterDivNk(wKeyIterDiv
                                NkInit),
                             .iNewKey(wNewKeyInit),
                             .oKeys(wKeys), .oKeysValid
                                (oKeysValid));

// round transformation blocks
Round R1(                    .iClk(iClk), .iReset
                                (iReset),
                             .iBlockIn(wBlockOutInit),
                             .iRoundKey(wRoundKey1),
                             .oBlockOut(wBlockOut1),
                             .iReady(wValidInit),
                             .oValid(wValid1));

Round R9(                    .iClk(iClk), .iReset
                                (iReset),
                             .iBlockIn(wBlockOut8),
                             .iRoundKey(wRoundKey9),
                             .oBlockOut(wBlockOut9),
                             .iReady(wValid8),
                              .oValid(wValid9));

// 10 rounds total
// Initial key addition
assign wRoundKeyFinal = wKeys[128*('Nr-7)-1:
   128*('Nr-8)];

// round key assignments
assign wRoundKey9     = wKeys[128*('Nr-6)-1: 128*('Nr-7)];
assign wRoundKey8     = wKeys[128*('Nr-5)-1: 128*('Nr-6)];
assign wRoundKey7     = wKeys[128*('Nr-4)-1: 128*('Nr-5)];
assign wRoundKey6     = wKeys[128*('Nr-3)-1: 128*('Nr-4)];
assign wRoundKey5     = wKeys[128*('Nr-2)-1: 128*('Nr-3)];
assign wRoundKey4     = wKeys[128*('Nr-1)-1: 128*('Nr-2)];
assign wRoundKey3     = wKeys[128*'Nr-1: 128*('Nr-1)];
assign wRoundKey2     = wKeys[128*('Nr+1)-1: 128*'Nr];

assign wRoundKey1     = wNkKeysInit[128-1:0];
assign wRoundKeyInit  = iKey[128-1:0];
```

```
FinalRound FinalRound(  .iClk(iClk), .iReset(iReset),
                        .iBlockIn(wBlockOut9),
                        .iRoundKey(wRoundKeyFinal),
                        .oBlockOut(oCiphertext),
                        .iReady(wValid9), .oValid
                        (oValid));
endmodule
```

Although pipelined designs such as the one above can potentially encrypt data at fast data rates, a problem arises in these architectures if one were to introduce new keys at a rate faster than the encryption speed. The surrounding system would have to be smart enough to wait for the pipe to empty before introducing the new data block along with the new key.

This information has to be fed back to the outside system that is providing the information and the corresponding keys so that they can be buffered and held appropriately. In the worst case where a new key is required for every block of data, the pipelined architecture would have a throughput equivalent to that of the iterative architecture and would be a massive waste of space (not to mention the disappointment of not achieving the advertised throughput). The next section presents an architecture that eliminates this problem.

4.1.7   Fully Pipelined Architecture

The term *fully pipelined* refers to an architecture for the key expansion that runs in parallel to the round transformation pipeline, where corresponding stages in the pipeline provide each other with the exact information at just the right time. In other words, the round key for any particular stage and any particular data block is valid for only one clock cycle and is used by the corresponding round at that time. This occurs in parallel for every pipeline stage. Thus, a unique key may be used for potentially every block of data, with no penalization in terms of latency or wait states. The maximum throughput of the round transformation pipeline is always achieved independent of the topology of the key set. A block diagram for the fully pipelined implementation is shown in Figure 4.10.

This means that a single iteration through the Key Expansion function (four 32-bit word expansion of the key) would happen fully synchronous with the round previous to the round that the key being generated would be used. Also, the latency for the Key Expansion block would have to maintain a clock latency equal to that of the Round block, typically equal to 1–4 clocks.

For the round key at any arbitrary key expansion block to arrive at its corresponding Round block appropriately, and on potentially every clock pulse, the timing must be very precise. Specifically, each key expansion block must generate a round key in exactly the same number of clock cycles that the Round block can generate its corresponding data. Also, the latency must be such that each key is valid when presented to the add-round-key sub-block. To handle these requirements, each key expansion block is divided into four incremental expansion

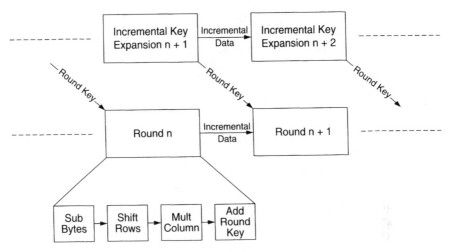

Figure 4.10   Fully pipelined key expansion.

blocks. Each incremental expansion block generates a single word ($128/4 = 32$ bits) for the key as described in the NIST specification. Each of these blocks is given a pipeline stage. This is shown in Figure 4.11.

As mentioned above, each Key-Exp1 block generates a single word (32 bits) of the expanded key. The stages to add pipelining is shown in Figure 4.12.

As can be seen in Figure 4.12, the S-box can be implemented as a synchronous 8×256 ROM, and to preserve latency timing accuracy, a pipeline stage must also be added to the R-CON calculation as described in the NIST specification.

Also, to ensure that the latency timing between the key pipeline and the data propagation pipeline is accurate, the keys must be generated one clock cycle earlier than the round data is completed. This is because the round key is necessary for the add-round-key block that clocks the XOR operation into its final register. In other words, clock 4 of the key expansion block must be synchronous with clock 3 of the corresponding Round block. This is handled by the initial key addition at the beginning of the key expansion process. This is shown in Figure 4.13.

The key data beyond the first 128 bits begins to expand on the first clock cycle, while the data pipeline begins on the second clock (a one-clock latency

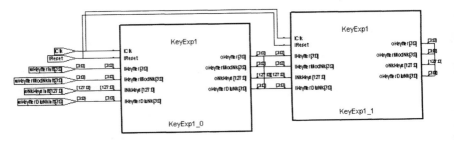

Figure 4.11   Propagation through 32-bit key expansion stages.

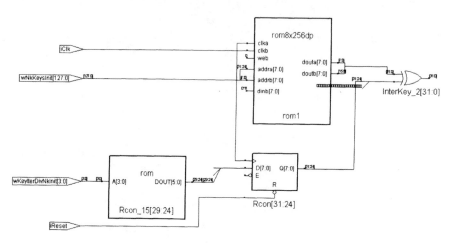

Figure 4.12   Single word expansion inside Key-Exp1.

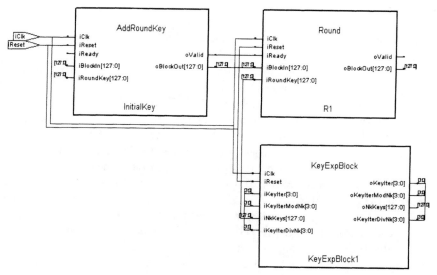

Figure 4.13   Skewing of the key pipeline.

from the initial add-round-key operation). The top-level implementation is shown in the code below for $N_k = 4$.

```
module AES_core(
  output [32*'Nb-1:0]    oCiphertext, // output cipthertext
  output                 oValid, // data at output is valid
  input                  iClk, iReset,
  input [32*'Nb-1:0]     iPlaintext, // input data to be
                             encrypted
  input [32*'Nk-1:0]     iKey, // input cipther key
  input                  iReady); // valid data to encrypt
```

```
wire [32*'Nb-1:0]     wRoundKey1, wRoundKey2, wRoundKey3,
                      wRoundKey4,
                      wRoundKey5, wRoundKey6, wRoundKey7,
                      wRoundKey8,
                      wRoundKey9, wRoundKeyFinal,
                      wRoundKeyInit;

wire [32*'Nb-1:0]     wBlockOut1, wBlockOut2, wBlockOut3,
                      wBlockOut4,
                      wBlockOut5, wBlockOut6, wBlockOut7,
                      wBlockOut8,
                      wBlockOut9, wBlockOutInit;

wire [32*'Nk-1:0]     wNkKeys1, wNkKeys2, wNkKeys3,
                      wNkKeys4,
                      wNkKeys5, wNkKeys6, wNkKeys7,
                      wNkKeys8,
                      wNkKeys9, wNkKeysFinal,
                      wNkKeysInit;

wire [3:0]            wKeyIter1, wKeyIter2, wKeyIter3,
                      wKeyIter4,
                      wKeyIter5, wKeyIter6, wKeyIter7,
                      wKeyIter8,
                      wKeyIter9, wKeyIterFinal,
                      wKeyIterInit;

wire [3:0]            wKeyIterModNk1, wKeyIterModNk2,
                      wKeyIterModNk3,
                      wKeyIterModNk4, wKeyIterModNk5,
                      wKeyIterModNk6,
                      wKeyIterModNk7, wKeyIterModNk8,
                      wKeyIterModNk9,
                      wKeyIterModNkFinal,
                      wKeyIterModNkInit;

wire [3:0]            wKeyIterDivNk1, wKeyIterDivNk2,
                      wKeyIterDivNk3,
                      wKeyIterDivNk4, wKeyIterDivNk5,
                      wKeyIterDivNk6,
                      wKeyIterDivNk7, wKeyIterDivNk8,
                      wKeyIterDivNk9,
                      wKeyIterDivNkFinal,
                      wKeyIterDivNkInit;

wire                  wValid1, wValid2, wValid3, wValid4,
                      wValid5, wValid6, wValid7, wValid8,
                      wValid9, wValidFinal, wValidInit;

// registered inputs
wire [32*'Nk-1:0]     wKeyReg;
wire                  wReadyReg;
wire [127:0]          wPlaintextReg;
```

```
// Initial key addition
assign wRoundKeyInit = wKeyReg[32*'Nk-1:32*'Nk-128];

// round key assignments
assign wRoundKey1    = wNkKeysInit[32*'Nb-1:0];
assign wRoundKey2    = wNkKeys1[32*'Nb-1:0];
assign wRoundKey3    = wNkKeys2[32*'Nb-1:0];
assign wRoundKey4    = wNkKeys3[32*'Nb-1:0];
assign wRoundKey5    = wNkKeys4[32*'Nb-1:0];
assign wRoundKey6    = wNkKeys5[32*'Nb-1:0];
assign wRoundKey7    = wNkKeys6[32*'Nb-1:0];
assign wRoundKey8    = wNkKeys7[32*'Nb-1:0];
assign wRoundKey9    = wNkKeys8[32*'Nb-1:0];

// register inputs
InputRegs InputRegs(    .iClk(iClk), .iReset(iReset),
                        .iKey(iKey),
                        .iPlaintext(iPlaintext),
                        .iReady(iReady), .oKey(wKeyReg),
                        .oPlaintext(wPlaintextReg),
                        .oReady(wReadyReg));

// initial key expansion
KeyExpInit KeyExpInit(  .iClk(iClk), .iReset(iReset),
                        .iNkKeys(wKeyReg),
                        .oKeyIter(wKeyIterInit),
                        .oKeyIterModNk(wKeyIterMod
                            NkInit),
                        .oNkKeys(wNkKeysInit),
                        .oKeyIterDivNk
                            (wKeyIterDivNkInit));

// initial addition of round key
AddRoundKey InitialKey( .iClk(iClk), .iReset(iReset),
                        .iBlockIn(wPlaintextReg),
                        .iRoundKey(wRoundKeyInit),
                        .oBlockOut(wBlockOutInit),
                        .iReady(wReadyReg),
                        .oValid(wValidInit));

// Number of rounds is a function of key size (10, 12, or
   14)

// Key expansion blocks
KeyExpBlock KeyExpBlock1(    .iClk(iClk), .iReset(iReset),
                        .iKeyIter(wKeyIterInit),
                        .iKeyIterModNk(wKeyIterMod
                            NkInit),
                        .iNkKeys(wNkKeysInit),
                        .iKeyIterDivNk(wKeyIterDiv
                            NkInit),
                        .oKeyIter(wKeyIter1),
```

```
                              .oKeyIterModNk(wKeyIter
                                ModNk1),
                              .oNkKeys(wNkKeys1),
                              .oKeyIterDivNk(wKeyIter
                                DivNk1));

KeyExpBlock KeyExpBlock8(     .iClk(iClk), .iReset(iReset),
                              .iKeyIter(wKeyIter7),
                              .iKeyIterModNk(wKeyIter
                                ModNk7),
                              .iNkKeys(wNkKeys7),
                              .iKeyIterDivNk(wKeyIter
                                DivNk7),
                              .oKeyIter(wKeyIter8),
                              .oKeyIterModNk(wKeyIter
                                ModNk8),
                              .oNkKeys(wNkKeys8),
                              .oKeyIterDivNk(wKeyIter
                                DivNk8));

// round transformation blocks

Round R1(                     .iClk(iClk), .iReset(iReset),
                              .iBlockIn(wBlockOutInit),
                              .iRoundKey(wRoundKey1),
                              .oBlockOut(wBlockOut1),
                              .iReady(wValidInit),
                               .oValid(wValid1));
                                  ...

Round R9(                     .iClk(iClk), .iReset(iReset),
                              .iBlockIn(wBlockOut8),
                              .iRoundKey(wRoundKey9),
                              .oBlockOut(wBlockOut9),
                              .iReady(wValid8),
                              .oValid(wValid9));

// 10 rounds total

assign wRoundKeyFinal = wNkKeys9[32*`Nb-1:0];

KeyExpBlock KeyExpBlock9(     .iClk(iClk), .iReset(iReset),
                              .iKeyIter(wKeyIter8),
                              .iKeyIterModNk
                                (wKeyIterModNk8),
                              .iNkKeys(wNkKeys8),
                              .iKeyIterDivNk
                                (wKeyIterDivNk8),
                              .oKeyIter(wKeyIter9),
                              .oKeyIterModNk
                                (wKeyIterModNk9),
                              .oNkKeys(wNkKeys9),
```

```
                                     .oKeyIterDivNk(wKeyIter
                                       DivNk9));
FinalRound FinalRound(               .iClk(iClk), .iReset(iReset),
                                     .iBlockIn(wBlockOut9),
                                     .iRoundKey(wRoundKeyFinal),
                                     .oBlockOut(oCiphertext),
                                     .iReady(wValid9), .oValid
                                       (oValid));
endmodule
```

4.2 PERFORMANCE VERSUS AREA

In this section, we discuss speed/area trade-offs for pipelined versus compact architectures and provide actual measurements from typical target technologies. All three architectures were designed with the same hardware description language (Verilog), and all used the same coding conventions.

The first target technology was a Xilinx Virtex II FPGA. The statistics are shown in Table 4.1.

For comparison against the FPGA implementation, the design was also targeted to an 0.35 μm ASIC process as shown in Table 4.2.

The performance metrics shown in the tables are defined as follows:

1. LUTs: This represents the logic utilization that the AES core consumes inside the FPGA.

Table 4.1 Speed/Area Statistics Targeting a Xilinx Virtex II

Architecture	Area (Xilinx LUTs)	Best Possible Throughput (MBPS)	Worst-Case Throughput
Iterative	886	340	340
Partially Pipelined	4432	15,400	314
Fully Pipelined	5894	15,400	15,400

Worst-case throughput assumes that a new key is introduced for every data block.

Table 4.2 Speed/Area Statistics for an 0.35-μm AMI ASIC

Architecture	Area (ASIC Gates)	Best Possible Throughput (MBPS)	Worst-Case Throughput
Iterative	3321	788	788
Partially Pipelined	15,191	40,064	817
Fully Pipelined	25,758	40,064	40,064

Worst-case throughput assumes that a new key is introduced for every data block.

2. ASIC gates: This is the number of logic gates that is consumed by the AES core in an ASIC.

3. Best possible throughput: This is the maximum number of data bits that can be processed per second in the best-case scenario. "Best case" refers to the situation where there is the least amount of penalty delay due to expanding new keys.

4. Worst-case throughput: "Worst case" here refers to the situation where there is the greatest amount of penalty delay due to expanding new keys. This situation arises when every data block has a unique key.

As can be seen from the data in Tables 4.1 and 4.2, the fully pipelined architecture is two orders of magnitude faster (in terms of throughput) under the worst-case scenario of a new key introduced for every data block. Note that the pipelined architecture is "penalized" for frequent key changes if it cannot fill the pipeline until previous encryptions are complete. This is what accounts for the drastic drop from best-case to worst-case throughput in the standard pipelined architecture.

4.3 OTHER OPTIMIZATIONS

As can be seen from the comparisons section, one of the primary issues with a fully pipelined design is area utilization. Also an issue with this pipelined architecture is the number of block RAMs needed to hold the S-box transformation look-up table. Many modern implementations use the block RAM LUT approach as it is easy to implement and can generate the necessary transformation in a single clock cycle. An arbitrary mapping over $GF(2^8)$ will require one 8×256 RAM module. A single mapping would not cause an issue, but considering that approximately 320 look-ups need to be performed on every clock, this presents a major issue for FPGA implementations because this memory requirement pushes the limits of even the larger modern FPGAs.

Another solution would be to implement the multiplicative inverse using the Extended Euclidean Algorithm as outlined in various technical references. However, the algorithm to compute the inverse of an arbitrary polynomial in $GF(2^m)$ has complexity of $O(m)$ and requires 2 m calculations. These steps cannot be placed in parallel for efficient hardware because each calculation depends on the previous. This type of iterative algorithm may be acceptable for software implementations, but the latency in a hardware implementation (2 * m * Rounds = 160 to 224 clocks) may be unacceptable for applications frequently encrypting small amounts of data.

A third approach to implementing the S-box in hardware was proposed by Vincent Rijmen (one of the inventors of Rijndael). The idea is to represent every element of GF(256) as a polynomial of the first degree with coefficients from GF(16). Denoting the irreducible polynomial as $x^2 + Ax + B$, the multiplicative inverse for an arbitrary polynomial bx + c is given by:

$$(bx + c)^{-1} = b(b2B + bcA + c2)^{-1}x + (c + bA)(b2B + bcA + c2)^{-1}$$

The problem of calculating the inverse in GF(256) is now translated to calculating the inverse in GF(16) and performing some arithmetic operations in GF(16). The inverse in GF(16) can be stored in a much smaller table relative to the mapping in GF(256) and will correspond with an even more compact S-box implementation.

Chapter 5

High-Level Design

Due to the increasing complexity of FPGA designs as well as the increasing sophistication and capability of the corresponding design tools, there will always be an increasing in demand for the ability to model a design with higher levels of abstraction. The move from schematic-based design to HDL (Hardware Description Language) design was revolutionary. It allowed a designer to describe modules in a behavioral fashion that was theoretically technology independent. There are many aspects of digital design with HDL that have become monotonous and time consuming and where higher levels of abstraction are obvious fits and have natural paths of progression. Other technologies that have been introduced over the past decade have had a difficult time migrating to mainstream because (and this point can be argued) the fit has been less obvious. This chapter discusses a number of high-level design techniques that have been found to provide utility to FPGA design engineers.

During the course of this chapter, we will discuss the following topics in detail:

- Abstract state-machine design using graphical tools
- DSP design using MATLAB and Synplify DSP
- Software/hardware codesign

5.1 ABSTRACT DESIGN TECHNIQUES

In the 17th century, the mathematician Leibniz postulated that a mathematical theory must be simpler than the system it describes, otherwise it has no use as a theory. This is a very profound statement and one that has a direct corollary to modern engineering. If a new form of abstract design is not simpler to comprehend or easier to design with than a previous form, then it is of little or no use to us in the real world. Some high-level design techniques that have been developed

Advanced FPGA Design. By Steve Kilts
Copyright © 2007 John Wiley & Sons, Inc.

69

over the past 10–15 years certainly fall under the category of "lateral abstraction," or new design techniques that are no more abstract than existing technologies. These technologies are slow to catch on; that is, if there is any benefit at all. This chapter discusses techniques that do provide the true benefits of abstraction and that can increase the effectiveness of a designer.

5.2 GRAPHICAL STATE MACHINES

One obvious fit for high-level design techniques is that of state-machine creation. Remembering back to our first course in logic design, we learned to design state machines by first drawing a state-transition diagram and then translating this to HDL (or gates as the course may be) manually. The state-diagram representation is a natural abstraction that fits well with our existing design process.

One feature of describing state machines in HDL is the vast number of valid state-machine encodings that are available to the designer. There are state-machine encodings that are optimal for compact designs and those that are well suited for high-speed designs. Many of the variations have to do with designer preference, but this also introduces the possibility of human error, not to mention software translation error when recognizing a state-machine description. Oftentimes a designer does not have the knowledge whether one state-machine encoding will be optimal for his or her implementation until after synthesis has been performed. Modifying the state-machine encoding is time consuming and is generally not related to the functionality of the state machine. It is true that most synthesis tools will recognize a state machine and recode it based on the physical design constraints, but a higher-level description that gives the synthesis tool the most flexibility regarding the implementation details is, in general, an optimal approach.

Take for example the following state machine that loads a multiplier and accumulator for a low-pass DSP function:

```
module shelflow(
  output reg         multstart,
  output reg [23:0]  multdat,
  output reg [23:0]  multcoeff,
  output reg         clearaccum,
  output reg [23:0]  U0,
  input              CLK, RESET,
  input      [23:0]  iData, // X[0]
  input              iWriteStrobe, // X[0] is valid
  input      [23:0]  iALow, iCLow, // coeffs for low pass filter
  input              multdone,
  input      [23:0]  accum);

  // define input/output samples
  reg        [23:0]  X0, X1, U1;
```

```
// the registers that are multiplied together in mult24
reg        [2:0] state; // holds state for sequencing
                        through mults

parameter        State0 = 0,
                 State1 = 1,
                 State2 = 2,
                 State3 = 3;

always @(posedge CLK)
if(!RESET) begin
  X0            <= 0;
  X1            <= 0;
  U0            <= 0;
  U1            <= 0;
  multstart     <= 0;
  multdat       <= 0;
  multcoeff     <= 0;
  state         <= State0;
  clearaccum    <= 0;
end
else begin
// do not process state machine if multiply is not done
case(state)
  State0: begin
    // idle state
    if(iWriteStrobe) begin
      // if a new sample has arrived
      // shift samples
      X0            <= iData;
      X1            <= X0;
      U1            <= U0;
      multdat       <= iData; // load mult
      multcoeff     <= iALow;
      multstart     <= 1;
      clearaccum    <= 1; // clear accum
      state         <= State1;
    end
    else begin
      multstart     <= 0;
      clearaccum    <= 0;
    end
  end
  State1: begin
    // A*X[0] is done, load A*X[1]
    if(multdone) begin
      multdat       <= X1;
      multcoeff     <= iALow;
```

```
      multstart   <= 1;
      state       <= State2;
    end
    else begin
      multstart   <= 0;
      clearaccum  <= 0;
    end
  end
  State2: begin
    // A*X[1] is done, load C*U[1]
    if(multdone) begin
      multdat     <= U1;
      multcoeff   <= iCLow;
      multstart   <= 1;
      state       <= State3;
    end
    else begin
      multstart   <= 0;
      clearaccum  <= 0;
    end
  end
  State3: begin
    // C*U[1] is done, load G*accum
    // [RL-1] U0 <= accumsum;
    if(multdone) begin
      U0              <= accum;
      state           <= State0;
    end
    else begin
      multstart   <= 0;
      clearaccum  <= 0;
    end
  end
  default
    state           <= State0;
  endcase
  end
  endmodule
```

The above module sequences through the operations for executing multiply and accumulate operations. For the purposes of this example, the multiplier and accumulator are located outside of this module. Let us take a look at the effect of describing this state machine in a more abstract format.

There are a number of tools that can be used for graphical state-machine entry. To begin with, let us use the free Xilinx state-machine editor called State-CAD. The graphical state machine is represented in Figure 5.1.

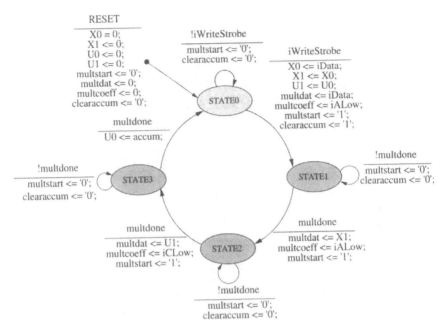

Figure 5.1   Graphical state machine design.

Clearly, the above representation is much easier to read than the Verilog representation because it represents the state machine as we would visualize it.

Graphical state machines are much easier to read and allow for automatic speed/area optimizations.

Despite the clarity of the graphical representation, however, the RTL that is generated is hardly more readable than a netlist. For brevity, we show only a few snippets (exported in the older Verilog-1995 format).

State Decoding:

```
if ( ~iWriteStrobe & STATE0 | multdone & STATE3 | RESET )
next_STATE0=1;
else next_STATE0=0;

if ( ~RESET & iWriteStrobe & STATE0 | ~RESET & ~multdone
    & STATE1)
next_STATE1=1;
else next_STATE1=0;

if ( ~RESET & multdone & STATE1 | ~RESET & ~multdone & STATE2 )
next_STATE2=1;
else next_STATE2=0;
```

```
if ( ~RESET & multdone & STATE2 | ~RESET & ~multdone & STATE3 )
next_STATE3=1;
else next_STATE3=0;
```

Output Decoding for one of the inputs to the multiplier:

```
multcoeff=({24{STATE0}}&({24{~RESET}}&({24{~RESET}}&{24
{iWRITESTROBE}}))&({iALow23, iALow22, iALow21, iALow20,
iALow19, iALow18, iALow17, iALow16, iALow15, iALow14,
iALow13, iALow12, iALow11, iALow10, iALow9, iALow8,
iALow7, iALow6, iALow5, iALow4, iALow3, iALow2, iALow1,
iALow0})) | ({24{STATE0}}&{24{~RESET}}&{24
{~iWriteStrobe}})&('h0)) | ({24{STATE1}}&{24{~RESET}}&
{24{~multdone}})&('h0)) | ({24{STATE1}}&({24{~RESET}}&
{24{multdone}})&({iALow23, iALow22, iALow21, iALow20,
iALow19, iALow18, iALow17, iALow16, iALow15, iALow14,
iALow13, iALow12, iALow11, iALow10, iALow9, iALow8,
iALow7, iALow6, iALow5, iALow4, iALow3, iALow2, iALow1,
iALow0})) | ({24{STATE2}}&{24{~RESET}}{24{~multdone}})&
({iCLow23, iCLow22, iCLow21, iCLow20, iCLow19, iCLow18,
iCLow17, iCLow16, iCLow15, iCLow14, iCLow13, iCLow12,
iCLow11, iCLow10, iCLow9, iCLow8, iCLow7, iCLow6, iCLow5,
iCLow4, iCLow3, iCLow2, iCLow1, iCLow0})) | ({24{STATE2}}&
{24{~RESET}}&{24{~multdone}})&('h0)) | ({24{STATE3}}&
{24{~RESET}}&{24{~multdone}})&('h0)) | ({24{STATE3}}&
{24{~RESET}}&{24{~multdone}})&('h0)) | (({24{RESET}})
&('h0));
```

When implemented, the two state machines show comparable results in terms of area utilization and performance. The state machine implemented in StateCad, however, can be optimized further for either performance or area based on the implementation options.

The downside to the StateCad implementation is that the autogenerated RTL is almost as unreadable as a netlist. Even if this is optimized, is it acceptable to have such unreadable code? The answer in most cases is, yes. When we made the transition from schematic design to RTL, we had to let go of the low-level representation and put more trust in our tools. This allowed us to design more complex circuits quicker, design at a more abstract level, produce source code that was portable between technologies, and so on. It is true that we analyze portions of the gate-level design to ensure proper logic implementation, but as tools become more and more refined and as we RTL designers become more experienced, this is required less frequently. Regardless, as a whole we can never expect to look at a synthesized netlist or sea of gates and derive any high-level meaning about a design. We must refer to the level of abstraction that the designer created directly to understand the design. In a similar fashion, when designing at levels of abstraction above RTL, we will not always have readable

RTL to reference. Instead, we will always have to rely on the following regardless of where abstract design takes us in the future:

1. We can analyze and understand the top level of abstraction. This is the level that the designer originally described the design.
2. We have tools to verify that lower levels of abstraction are implemented as we intended.

Of key importance is the readability of the top level of abstraction where the design takes place. Of less importance is the readability of the autogenerated RTL.

There are still FPGA designers out there that design with schematics and have not been able to make the move to RTL-based design. It is easy to become comfortable with a methodology and feel that it is unnecessary to learn the latest design methods. This is a dangerous attitude, as it will cause a designer to lose effectiveness as an engineer or even become obsolete.

5.3 DSP DESIGN

DSP design is another great fit for abstract design techniques because design of a DSP with any reasonable complexity already takes place at a higher level of abstraction. DSP design is very mathematical by nature, and thus design tools such as MATLAB are commonly used for the development and analysis of these algorithms. Tools like MATLAB make it very easy to construct DSPs out of pre-existing building blocks and provide sophisticated tools to analyze frequency response, phase response, distortion characteristics, and many of the other metrics associated with digital filters.

Traditionally, the DSP algorithms are designed and analyzed with these high-level mathematical tools, and once the design is verified, the designer must convert these algorithms to synthesizable constructs manually for the FPGA target. The bridge that is being gapped by abstract FPGA design techniques is to take these preexisting highlevel descriptions and convert them directly to FPGA implementations. This is truly an abstract approach that simplifies the design process and allows the designer to focus on the top level of abstraction.

One tool that does a particularly good job of this is Synplify DSP from Synplicity. The Synplify DSP tool runs as an application within MATLAB, allowing a close coupling between MATLAB constructs and modeling capabilities with Synplicity's DSP to RTL synthesis. The system represented in Figure 5.2 is a Simulink model of a basic FIR filter.

The "Port In" and "Port Out" blocks indicate the I/O of the FPGA, and the FIR block is the DSP function to be synthesized to the FPGA. The Sinewave and Scope blocks are used only for MATLAB/Simulink simulation. Finally, the FDA Tool block is used to parameterize the FIR filter as shown in Figure 5.3.

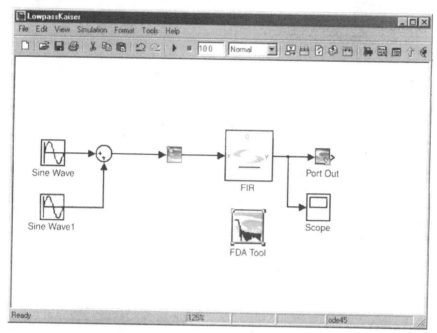

Figure 5.2   Simulink model of an FIR filter.

In Figure 5.3, a low-pass FIR filter is parameterized with a pass frequency of 0.1 (normalized to the sampling frequency). Any one of the filter characteristics can be modified with the corresponding response shown in the adjacent window. Once the parameters are set, a MATLAB/Simulink simulation can be run to verify the filter. Finally, Synplify DSP will generate the following Verilog code for the filter (exported with mixed Verilog-1995 and Verilog-2001 formats):

```
// AUTO-GENERATED CODE FROM SYNPLIFY DSP
module FIR( clk, gReset, gEnable, rst, en, inp, outp);
parameter inpBitWidth = 16;
parameter inpFrac = 8;
parameter coefBitWidth = 10;
parameter coefFrac = 8;
parameter dpBitWidth = 17;
parameter dpFrac = 8;
parameter outBitWidth = 17;
parameter tapLen = 46;
parameter extraLatency = 0;
input clk;
input gReset;
input gEnable;
input rst;
input en;
```

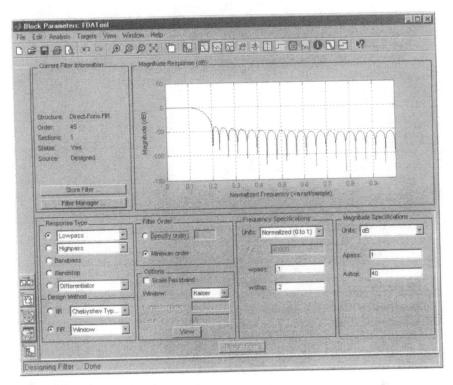

Figure 5.3   FIR parameterization.

```
input [inpBitWidth-1:0] inp;
output [outBitWidth-1:0] outp;

wire signed [coefBitWidth-1:0] CoefArr [0:tapLen + 0 - 1];
generate
begin: CoefArrGen
  assign CoefArr[0] = 10'b0000000000;

  ...
  assign CoefArr[45] = 10'b0000000000;
end
endgenerate
wire signed [inpBitWidth-1:0] multInp;
wire signed [coefBitWidth + inpBitWidth-1:0] multOuts
            [0:tapLen-1];
wire signed [coefBitWidth + inpBitWidth-1:0] multBufs1
            [0:tapLen-1];
wire signed [coefBitWidth + inpBitWidth-1:0] multBufs1pre
            [0:tapLen-1];
wire rstBuf;
wire enBuf;
reg signed [dpBitWidth-1:0] mem [0:tapLen-2];
```

```
assign multInp = inp;
generate
 genvar i1;
 for(i1=0; i1<=tapLen-1; i1=i1+1)
 begin: multOuts_gen
    assign multOuts[i1] = multBufs1[i1];
 end
endgenerate
assign rstBuf = rst;
assign enBuf = en;
assign outp = multOuts[0] + mem[0];
generate
 genvar i2;
 for(i2=0; i2<=tapLen-1; i2=i2+1) begin: floop
    assign multBufs1pre[i2] = multInp * CoefArr[i2];
    assign multBufs1[i2] = multBufs1pre[i2] >>>
                                (coefFrac+inpFrac-dpFrac);
 end
endgenerate

generate
 integer i3;
 begin: inner_floop
    always @(posedge clk) begin
      if( (rstBuf==1) || (gReset==1) ) begin
         for(i3 = 0; i3 <= tapLen-2; i3=i3+1) begin
            mem[i3] <= 0;
         end
         end // reset
      else if( (enBuf==1) && (gEnable==1) ) begin
         for(i3 = 0; i3 <= tapLen-3; i3=i3+1) begin
            mem[i3] <= mem[i3+1] + multOuts[i3+1];
         end
            mem[tapLen-2] <= multOuts[tapLen-1];
      end // enable
    end // always
 end // inner_floop
endgenerate
endmodule
```

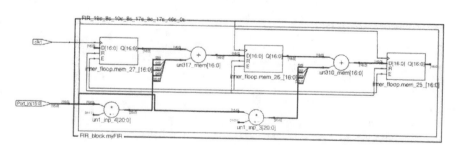

Figure 5.4   Auto-Generated Pipelined FIR.

Table 5.1   Implementation Results for Pipelined FIR

Register	806
LUT	828
Speed	140 MHz

The code generated above assumes a pipelined architecture where a new sample can be applied on every clock edge. The pipelined architecture is shown in Figure 5.4.

As can be seen in Figure 5.4, the input sample is multiplied by the coefficients and added to the appropriate stage in the pipeline. The above topology is consistent for all taps of the FIR. The final implementation results are listed in Table 5.1.

Thus, with a 16-bit data path, the FIR is capable of handling data at a rate of 2.24 Gbps (2.24×10^9 bits/second). The primary drawback to the above architecture is that the amount of space required in an FPGA implementation is relatively large. In many DSP applications, there are a certain number of clocks per sample (i.e., the system clock frequency is greater than the sampling frequency), which means that a more compact architecture may be used to reuse the same DSP hardware for the required MAC operations. With an abstract design tool such as Synplify DSP, this architectural modification can be made as an implementation option during MATLAB to RTL synthesis. For example, Synplify DSP provides an option for "folding," which essentially folds up the pipeline to reuse hardware resources. The amount of folding is dependent on the number of clocks available per sample period. In general, the worst-case ratio between the slowest clock period and the fastest sample rate will provide a guide to the maximum amount of folding possible. For example, if we take the above FIR implementation and define 200 clocks per sample period, there is enough time to calculate 45 taps with a single set of MAC hardware. This is shown in Figure 5.5.

In the architecture of Figure 5.5, all filtered samples for a 45th order filter are queued in the output shift register. When a new sample arrives, the ROM begins sequencing to multiply the input sample by every coefficient. The shift register holding all filtered samples begins shifting and adding the input sample multiplied by the appropriate coefficient. At the end of the sequence, all output samples have been shifted to the appropriate location with the addition of the scaled version of

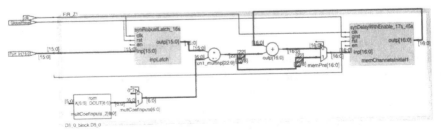

Figure 5.5   FIR logic folded for area efficiency.

Table 5.2　Implementation Results for Folded FIR

Registers	938
LUT	249
Speed	120 MHz

the input sample. This is a folded version of the pipelined architecture and is much more compact as shown in the resource report listed in Table 5.2.

The total area has been reduced dramatically with the compact architecture, with a trade-off in the reduction of maximum throughput that has now been reduced to one sample every 45 clocks, or 42 Mbps (42×10^6 bits/second).

Some abstract design tools such as Synplify DSP allow for automatic architectural trade-offs such as pipelined versus folded implementations.

5.4　SOFTWARE/HARDWARE CODESIGN

Looking back at our definition of higher (and better) levels of abstraction, C-based FPGA design is one that has traditionally sat very close to the border. In the 1990s, a number of companies aggressively pursued C-level synthesis as a replacement for HDL design. Some companies were advertising that their tool could convert ANSI-compliant C code to synthesizable HDL. The problem with this claim is that C code (and software design for microprocessors in general) is a sequential, instruction-based language. One instruction is executed after another in a sequence. To convert this to synthesizable HDL, this sequence of commands had to be converted to a state-machine format, and for any design of any reasonable size, this produced code that was unusable.

To remedy this problem, these companies developed various flavors of cycle-based C syntax that would convert easily to HDL. The fundamental problem with this approach was that a new HDL language had been invented that was no more easy to use than the standard languages for HDL. Additionally, they required an extra synthesis step (sometimes automated, sometimes not) that could have been easily bypassed in the first place. The argument in favor of the cycle-based C language was that it could be compiled and run against other C models, but for the majority of HDL designers, this was unnecessary for most of their applications. Although there are applications where this is a fit, C-based design still has not caught on in the mainstream FPGA world because the EDA companies have not convinced the design community that there is a benefit or simplification of the design process with C-level design.

One of the primary benefits of C-level design today is the ability to simulate hardware and software in the same environment. Despite this, it is still important to understand where to divide the line between software and hardware implementation. A few of the issues to consider are listed here.

Complexity: There are certain algorithms that simply lend themselves well to a software implementation over hardware. Sophisticated algorithms that require recursion or a number of iterations that is not well defined prior to executing the code will usually be better suited for software. An example may include a successive approximation algorithm where certain error functions are monitored to determine an acceptable stopping point (discussed in Chapter 8). Another example would be floating-point operations. A simple floating-point calculation can easily be performed in hardware if the boundary conditions are understood up front, but to be truly IEEE compliant, a floating-point coprocessor is much more efficient than a dedicated hardware implementation.

Speed: Any operations that need to run extremely fast are usually targeted toward the hardware implementation. There will always be a certain amount of overhead when running an algorithm in software and associated delays due to other unrelated events. Hardware designs, on the other hand, can be highly optimized for high-throughput or low-latency timing requirements (see Chapter 1).

Repetition: A task that is repeated continuously, even if it is not complex, could have the effect of slowing down the normal operation of the microprocessor. Thus, a microprocessor can often benefit from off-loading very repetitive tasks to dedicated hardware. Examples may include a function that scans for events or any type of continuously modulated output.

Real-Time Precision: A microprocessor executes instructions in a specific order, and any tasks will have to wait to be serviced. The amount of wait time depends on the priority of the triggering event, but often too many high-priority events begin to step on each other and the microprocessor cannot service these events fast enough. Additionally, if a function must be timing accurate down to a single clock cycle, only a hardware implementation will be able to guarantee this level of precision.

Operating System or User Interface: If any type of operating system is required, then a microprocessor (most likely 32 bits) will be required to support it. Also, if even a simple user interface is required over a common bus format, it is often easier to use an embedded 8-bit microprocessor with a predefined peripheral rather than design something from scratch (this includes bit-banging around the peripheral itself).

Many of these decisions cannot be made by a machine and depend on the judgment of the designer. Understanding the criteria for partitioning a design between hardware and software will be a requirement of system engineers into the foreseeable future.

5.5 SUMMARY OF KEY POINTS

- Graphical state machines are much easier to read and allow for automatic speed/area optimizations.
- Of key importance is the readability of the top level of abstraction where the design takes place. Of less importance is the readability of the autogenerated RTL.
- Some abstract design tools such as Synplify DSP allow for automatic architectural trade-offs such as pipelined versus folded implementations.

Chapter 6

Clock Domains

Interestingly, a number of traditional textbooks on the topic of digital design, particularly when referring to FPGAs, take a hard-line stance on using only one clock domain for an entire design. In other words, only a single net may drive the clock inputs of all flip-flops in a design. Although this would greatly simplify timing analysis and eliminate many of the problems associated with multiple clock domains, the use of only one clock is often not possible due to various system constraints outside the domain of the FPGA. Oftentimes, FPGAs are used to pass data between two systems with predefined clock frequencies, receive and transmit data over multiple I/O interfaces, process asynchronous signals, and to prototype low-power ASICs with gated clocks. The attempt of this chapter is to provide practical guidance relative to the problems and solutions associated with multiple clock domains and asynchronous signals in FPGA designs.

A clock domain, as it is referred to here and in following chapters, is a section of logic where all synchronous elements (flip-flops, synchronous RAM blocks, pipelined multipliers, etc) are clocked by the same net. If all flip-flops are clocked by a global net, say the main clock input to the FPGA, then there is one clock domain. If there are two clock inputs to the design, say one for "interface 1" and one for "interface 2" as shown in Figure 6.1, then there are two clock domains.

Gated clocks, derived clocks, and event-driven flip-flops all fall into the category of clock domains. Figure 6.2 illustrates the creation of a new clock domain through a simple gated clock. Note that this type of clock control is not recommended for FPGA designs (one would typically use a clock-enable input on the second bank of flip-flops), but it serves to illustrate the concept.

During the course of this chapter, we will discuss the following topics in detail:

- Passing signals between two different clock domains.

    Cause of metastability and the effect it can have on design reliability

    Avoiding metastability through phase control

    Double flopping to pass individual signals between clock domains

Advanced FPGA Design. By Steve Kilts
Copyright © 2007 John Wiley & Sons, Inc.

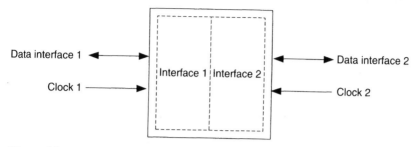

Figure 6.1 Dual clock domains.

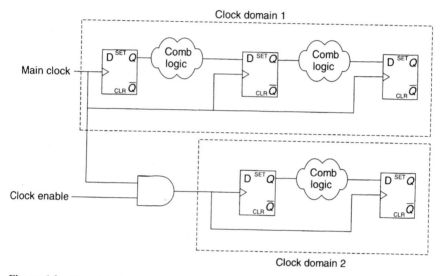

Figure 6.2 Clock domain creation via gated clock.

FIFOs for passing multibit words between clock domains
Partitioning synchronizer blocks to improve design organization

- Handling gated clocks in ASIC prototypes.

Establishing a single clocks module
Automatic gating removal

6.1 CROSSING CLOCK DOMAINS

The first problem that needs to be addressed when dealing with multiple clock domains is the issue of passing signals between the domains. There are a number of reasons clock domain crossing can be a major problem:

1. The failures are not always repeatable. If you have two clock domains that are asynchronous, then failures are often related to the relative timing between the clock edges. The clock often comes from an external source not in any way related to the actual functionality of your device.

2. Problems will vary from technology to technology. Often one will find that higher speed technologies with smaller setup and hold constraints will have statistically fewer problems than slower technologies (although due to other effects this is not always the case). Also, factors such as the implementation of the synchronous device, such as the manner in which the output is buffered, will also have a significant impact on the probability of a failure.

3. EDA tools typically do not detect and flag these problems. Static timing analysis tools analyze timing based on individual clock zones and will only perform interclock analysis if they are specified to do so in a specific manner.

4. In general, cross-clock domain failures are difficult to detect and debug if they are not understood. It is very important that all interclock interfaces are well defined and handled before any implementation takes place.

Let us first discuss what can go wrong when passing signals between clock domains. Consider the situation in Figure 6.3 where a signal is passed between two clock domains.

As shown in Figure 6.4, the slow clock domain has exactly twice the period of the fast clock domain. The time from the rising edge of the slow clock to the rising edge of the fast clock is always constant and equal to dC. Due to the matched phases of these clocks, dC will always remain constant (assuming no frequency drift) and in this case is always greater than the logic delay plus the setup time of the flip-flop clocked by "fast clock."

When these clocks started up, they had a phase relationship that avoided any setup or hold timing violations. As long as neither clock drifts, no timing violations will occur, and the device will work as expected. Now consider the scenario where the same clocks power up with the phase relationship as shown in Figure 6.5.

In this scenario, the clock edges line up to create a timing violation. This scenario can occur between any two clock domains of any relative frequencies. However, if the frequencies are not well matched, the violations will not occur in such a regular pattern.

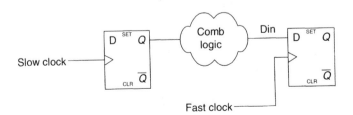

Figure 6.3   Simple propagation between clock domains.

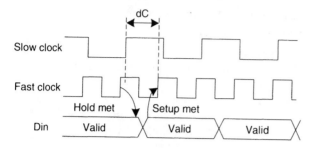

Figure 6.4 Timing between clock domains.

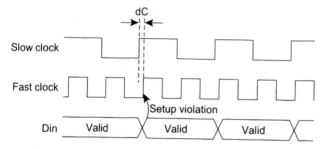

Figure 6.5 Phase relationship creates timing violation.

Clock synchronization issues are generally not repeatable and will affect the reliability of the FPGA design.

Solutions to these problems are discussed later in this chapter, but first we need to discuss what really happens when setup and hold violations occur. This topic is covered in the next section.

6.1.1 Metastability

A timing violation occurs when the data input to a flip-flop transitions within a window around the active clock edge as defined by the setup and hold times. This timing violation exists because if the setup and hold times are violated, a node within the flip-flop (an internal node or one that is exposed to the outside world) can become suspended at a voltage that is not valid for either a logic-0 or logic-1. In other words, if the data is captured within the window described above, the transistors in the flip-flop cannot be reliably set to a voltage representing logic-0 or logic-1. Rather than saturating at a high or low voltage, the transistors may dwell at an intermediate voltage before settling on a valid level (which may or may not be the correct level). This is called metastability and is illustrated in Figure 6.6.

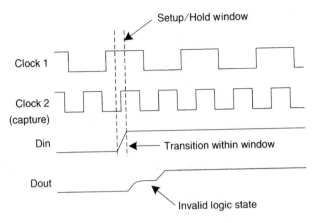

Figure 6.6   Metastability caused by timing violation.

As can be seen in the waveforms, a transition within the boundaries of the setup and hold conditions means that the output could rise to a voltage level that is not valid for either logic value. If a flip-flop contains an output buffer, the metastability may manifest itself as a spurious transition at the output as the internal signal settles. The amount of time the output can stay metastable is probabilistic, and it is possible for the output to remain metastable for the entire clock period. Thus, if this metastable value is fed into combinatorial logic, incorrect operations may occur depending on the thresholds of the logic gates. From a timing closure standpoint, the delay through the logic from one flip-flop to another is assumed to be less than the minimum clock period, but with a metastable signal the duration of metastability will consume available path delay. Clearly, a metastable signal can cause catastrophic functional failures in a design and will be very inconsistent depending on the relationship of the clock edges.

Metastability can cause catastrophic failures in the FPGA.

One important thing to note about the FPGA design flow is that simulating the effects of metastability can be very difficult. Digital-only simulators will not check for setup and hold violations, and then propagate a logic-X (unknown) if the violation occurs. In an RTL simulation, no setup and hold violations occur, and thus no signal will ever go metastable. Even with gate-level simulations, that check for setup and hold violations, it may be a difficult matter to simulate a condition where two asynchronous signals line up to cause a synchronization failure. This is especially difficult when the design or verification engineer is not looking for the problem in the first place. Thus, it is extremely important to understand how to design for reliability and avoid the need to uncover synchronization issues in simulation. There are a number of solutions to this problem, and these are discussed in the remainder of this chapter.

6.1.2 Solution 1: Phase Control

Consider two clock domains of different periods and with an arbitrary phase relationship. If at least one of the clocks is controllable inside the FPGA via an internal PLL (Phase locked loop) or DLL (Delay locked loop) and one of the clocks has a period that is a multiple of the other within the resolution of the PLL or DLL, then phase matching can be used to eliminate timing violations as shown below.

Consider the example where a signal is passed from a slow clock domain to a domain with half the period. Without any guarantee of the phase relationship between the clocks, timing violations may occur as described above. However, by using a DLL to derive the faster clock from the first, phase matching can be achieved.

In Figure 6.7, the DLL adjusts the phase of the faster (capture) clock domain to match that of the slower (transmitting) clock domain. The total amount of time available for data to pass between the two domains dC is always at its maximum possible value. In this case, as long as the propagation delay between the flip-flop from the slow register to the flip-flop of the fast register is less than the period of the fast clock, no setup violations will occur. If the skew cannot be tightly matched in a way to ensure hold time compliance, the fast could also be configured to capture the signal on the falling edge assuming there is enough slack to maintain setup time compliance.

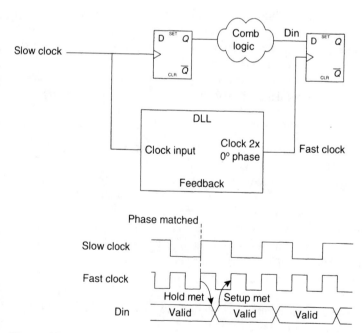

Figure 6.7 DLL for phase matching.

The phase-control technique can be used when the period of one clock is a multiple of the other and when one of the clocks can be controlled by an internal PLL or DLL.

In many cases, the designer does not have the luxury of controlling the phase relationships between clock domains. In particular, this occurs when specific timing requirements are imposed on the FPGA from outside the chip or if the periods of the two clock domains have no relationship to each other. For instance, if the FPGA is providing an interface between two systems that have very tight timing requirements imposed on the input and output delays of the chip, adjusting the phase of either clock may not be possible. Cases such as these arise very frequently, and new methods must be derived for dealing with them. The most common techniques are discussed in the following sections.

6.1.3 Solution 2: Double Flopping

Double flopping is a technique that can be used when passing single-bit signals between two asynchronous clock domains. As discussed in previous sections, a setup or hold violation may cause a node within a flip-flop to become metastable, and there will be an undefined amount of dwell time before the signal settles at a valid level. This dwell time adds to the clock-to-out time (and subsequently to the propagation delay of the path) and may cause a timing violation on the next stage. This is particularly dangerous if the signal feeds into a control branch or a decision tree. Unfortunately, there is no good way of predicting how long the metastability will last, nor is there a good way of back-annotating this information into the timing analysis and optimization tools. Assuming the two clock domains are completely asynchronous (phase control is not possible), a very simple way to minimize the probability of metastability is to use double-flopping. Note that other texts may refer to these as synchronization bits, dual rank flip-flops, or dual rank synchronizers.

In the configuration shown in Figure 6.8, the first flop in the synchronizer circuit (with the input labeled Din) may experience metastability but will have a chance to settle before it is relatched by the second stage and before it is seen by other logic structures. This is illustrated in Figure 6.9.

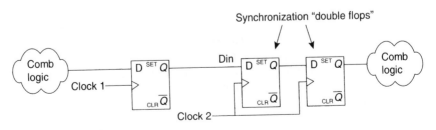

Figure 6.8 Double flopping.

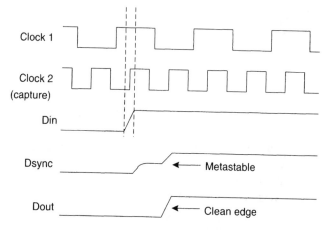

Figure 6.9 Double-flop resynchronization.

Here, Dsync is the first flip-flop, and Dout is the output of the second flip-flop. Dout essentially protects any other circuitry from seeing the metastable signal and passes it on once the synchronized signal has had a chance to settle. By adding no logic between the two flip-flops, we maximize the amount of time provided to the signal to settle.

Double flopping can be used to resynchronize single-bit signals between two asynchronous clock domains.

In theory, an output could remain metastable indefinitely, but in reality it will settle due to higher order effects of a real system. As an illustration, think of a ball parked perfectly at the top of a hill. A small nudge in either direction will send the ball down one side or the other. Likewise with a metastable logic gate, the random fluctuations in heat, radiation, and so forth, will push the metastable output to one state or another.

When sampling an asynchronous signal using the double-flopping technique, it is not possible to fully predict whether the desired transition will occur on the clock you expected or the following clock. This is usually not helpful when the bits are part of a bus containing data (some of the bits may transition a clock later than its companions) or when the arrival of the data is critical down to the precision of a single clock. It is useful, however, when used for control signals that can withstand a variance of ± 1 clock or more.

For instance, an external event that toggles a bit to trigger an action inside the FPGA may happen slowly and can handle a delayed reaction time of microseconds or even milliseconds. In this case, a few additional nanoseconds will not affect the behavior. If the bit driven by the external event feeds into the control structure of a state machine, the desired transition may be delayed by one clock cycle using double flopping. However if double flopping is not used, different portions of the decision logic may interpret the metastable state differently and activate multiple branches in the state machine simultaneously!

In addition to digital-only systems, a common scenario is that of a mixed-signal system that generates asynchronous feedback signals to the FPGA. This is illustrated in Figure 6.10.

The Verilog code to implement double flopping on an asynchronous signal is trivial:

```
module analog_interface(
    ...
    output reg fbr2,
    input   feedback);
    reg     fbr1;

    always @(posedge clk) begin
        fbr1 <= feedback;
        fbr2 <= fbr1; //; double flop
    end
    ...
```

The signal feedback may cause timing violations, and fbr1 may be metastable for an undetermined time after the clock edge. Therefore, fbr2 is the only usable signal for other logic.

When using the double-flopping method, it is important to specify timing constraints such that the signal path between the first and second clock domain is ignored during timing analysis. Because the double-flop structure resynchronizes the signal, there is no valid synchronous path between the two domains. In addition, the timing between the flip-flops should be minimized to reduce the probability that the metastable state is propagated through the second flip-flop.

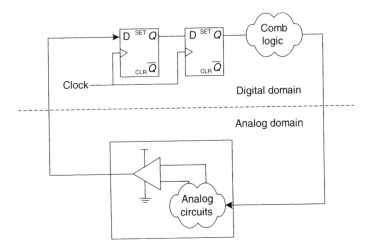

Figure 6.10   Resynchronization of analog feedback.

Timing analysis should ignore the first resynchronization flip-flop and ensure that the timing between the synchronization flip-flops themselves is minimized.

6.1.4 Solution 3: FIFO Structure

A more sophisticated way of passing data between clock domains is through the use of a first-in, first-out (FIFO) structure. FIFOs can be used when passing multi-bit signals between asynchronous clock domains. Very common applications for FIFOs include passing data between standardized bus interfaces and reading/writing burstable memory. For example, Figure 6.11 illustrates the interface between burstable memory and a PCI bus.

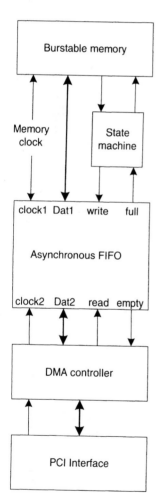

Figure 6.11   FIFO in a PCI Application.

FIFOs can be used when passing multibit signals between asynchronous clock domains.

FIFOs are very useful data structures for a variety of applications, but in the context of this discussion we will be concerned with its ability to handle bursts of data that need to be passed between clock domains.

The simplest analogy for a FIFO is a line at the supermarket. Customers arrive at the checkout at more or less random times and at a particular average frequency. Sometimes there will be little traffic, and at other times there will be bursts of customers. The cashiers at the checkout cannot immediately service every customer as they arrive, and thus a line forms. In an abstract sense, a line of data is called a queue. Subsequently, the cashiers proceed to service the customers at a more or less constant frequency regardless of the length of the line. If the average rate of customers exceeds the rate at which they can be serviced, the structure will be unsustainable. At that point, either another mechanism must be put in place to service the customers at a faster rate or the rate of new customers must decrease.

The same principles hold for many types of data transfers. Data may arrive on one clock domain at essentially random time intervals, some of which may contain large bursts of traffic. The receiving device, in this case sitting on a different clock domain, can only process the data at a particular rate. The queue that is formed takes place inside a device called a FIFO as shown in Figure 6.12.

With an asynchronous FIFO, data can arrive at arbitrary time intervals on the transmission side, and the receiving side pulls data out of the queue as it has the bandwidth to process it. Due to the finite size of any queue implemented with a FIFO, certain controls need to be in place to prevent an overflow. Two options are available for this scenario:

- Prior knowledge of the transmission rate (burstable or nonburstable), the minimum receiving rate, and the corresponding maximum queue size.

- Handshaking controls.

Note that it is not necessary for the clock domain of the transmission device to run at a faster frequency than the clock domain of the receiving device for an overflow to occur. A slower clock domain may require fewer clock cycles to pass

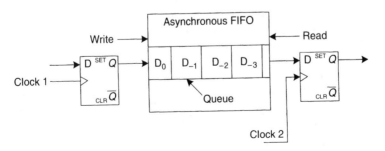

Figure 6.12   Asynchronous FIFO.

data to the FIFO than the number of clock cycles for the receiving side to process the data. Thus, if handshaking controls are not in place, it is critically important to understand the worst-case scenario as described above.

Note that if the transmitting side passes data to the FIFO at a rate faster than the receiving side can handle it for an arbitrary amount of time, the system simply will become unsustainable as the queue increases indefinitely. Because no memory device can store an unlimited amount of data, this issue would need to be addressed at the system architecture level. In general, the transmissions will arrive in bursts separated by periods of little or no activity. The maximum FIFO size would then need to be equal to or greater than (depending on the properties of the receiver) the size of the burst.

In many cases, neither the size of the bursts nor the distribution of the arriving data can be well defined. In this case, handshaking controls are necessary to control the data flow into the FIFO. This is often implemented with flags as shown in Figure 6.13. These include a full flag to inform the transmitting side that there is no more room in the FIFO and an empty flag to inform the receiving side that there is no more data to fetch. A state machine may be needed to manage the handshaking controls as illustrated in Figure 6.13.

FIFOs in FPGAs are typically implemented with a wrapper around a dual-port RAM. The seemingly trivial flags such as full and empty are in reality the difficult features to implement. The reason is because the flags for the input controls are often generated by the output stage, and similarly the flags for the output controls are often generated by the input stage. For instance, the logic that drives the input data must know whether or not the FIFO is full. This can only be determined by the amount of data that has been read by the output stage. Likewise, the logic that reads the data at the output stage must know if there is any new data available (whether the FIFO is empty). This can only be determined based on the write pointer from the input stage.

The purpose of the FIFO in this context is to handle the data transfer between asynchronous clock domains, but in the implementation of the FIFO itself we run into the same problems with respect to the handshaking flags. To pass the necessary signals from one domain to the other, we must revert to a technique such as double flopping as discussed in the previous section. Consider the diagram of a simplified asynchronous FIFO as shown in Figure 6.14.

In Figure 6.14, both the write address and read address must be resynchronized when passed to the other domain for empty and full generation. The problem that arises is that during the resynchronization of a multibit address, some of the bits may lag others by a clock cycle, depending on the individual propagation times of the individual traces. In other words, due to the asynchronous nature of the two clock domains, some bits may be captured on one edge of the capture clock, and the others may be captured on the next depending on whether the data arrives at the first flip-flop with sufficient time prior to the clock edge. This could be disastrous if some of the bits of a binary address change and others do not, as the receiving logic will see a completely invalid address not equal to either the previous or current address.

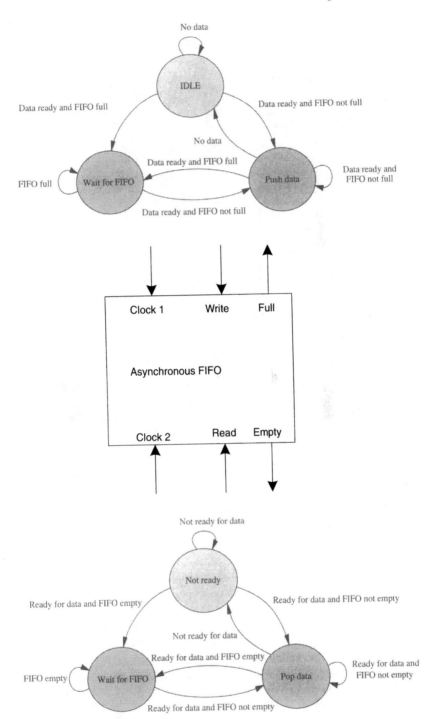

Figure 6.13 FIFO handshaking.

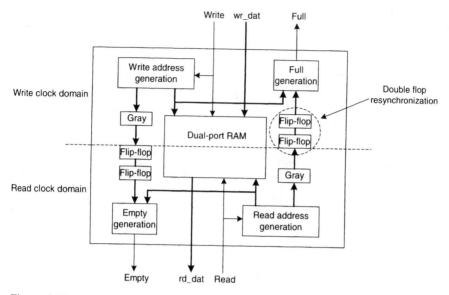

Figure 6.14   Simplified asynchronous FIFO.

This problem is solved by converting the binary address to gray code. A gray code is a special counter where adjacent addresses differ by only one bit. If every change to the address toggles only one bit, this eliminates the problem discussed above. If the one bit that changes is not captured by the next clock edge, the old address will remain as the synchronized value. Thus, any possibility for an incorrect address (something other than the old address and the current address) is eliminated.

Gray codes can be used to pass multibit counter data between asynchronous clock domains and are often used inside FIFOs.

An additional point is that just because the addresses that are passed through the asynchronous boundaries may arrive one clock cycle late does not necessarily mean that empty or full flags will be asserted incorrectly causing an overflow condition. The worst-case scenario is that the address is delayed. If this occurs when transferring the address to the read domain, the read logic will simply not realize that data has been written and will assume an empty condition when there is none. This will have a small impact on the overall throughput but will not cause an underflow (read when empty) condition. Similarly with the data that is passed into the write domain, if the read address is delayed, the write logic will assume that there is no space to write even though the space exists. This will also have a small impact on overall throughput but will not cause an overflow (write when full).

FIFOs are common enough that most FPGA vendors provide tools that automatically generate soft cores based on specifications from the user. These custom FIFOs can then be manually instantiated in the design similar to other blocks of IP (Intellectual Property). Thus, it is very likely these issues will not have to be

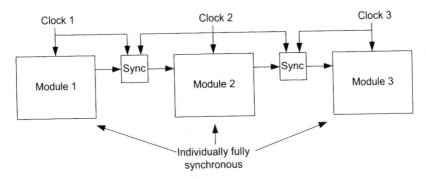

Figure 6.15   Partitioned synchronizer blocks.

addressed by the designer for a specific FIFO implementation in an FPGA. Similar issues, however, arise very frequently when passing data between domains, and a good understanding of these design practices are important for the advanced FPGA designer.

6.1.5   Partitioning Synchronizer Blocks

As a matter of good design practice, the top-level design should be partitioned such that the synchronizer blocks are contained in individual modules outside of any functional blocks. This will help to achieve the ideal clock domain scenario (one clock for the entire design) on a block by block basis. This is illustrated in Figure 6.15.

This is good partitioning for a number of reasons. First, the timing analysis on each functional block becomes trivial because it is fully synchronous. Second, the timing exceptions are easily definable when they apply to the entire sync block. Third, the synchronizers along with the corresponding timing exceptions are brought out to the top level lowering the probability that one will be over-looked due to human error.

> Synchronization registers should be partitioned as independent blocks outside of the functional modules.

There are a number of good design practices similar to this that apply when designing ASICs to be targeted at an FPGA prototype. This is discussed in the next section.

6.2   GATED CLOCKS IN ASIC PROTOTYPES

ASIC designs often have very tight requirements for power dissipation, and because of the flexibility in the design of an ASIC clock tree, it is very common for gated clocks to be scattered throughout the chip to disable any activity when it is not required. Although the FPGA prototype for this ASIC will be able to emulate the logical functionality, it will not have the same physical characteristics such as power dissipation. Thus, it is not necessarily a requirement that the FPGA emulate all low-power optimizations of the ASIC. In fact, due to the coarse nature of an FPGA's clock resources, it is not always possible (or desirable) to emulate this functionality. This section discusses methods to handle these situations and discusses techniques that can be applied to the ASIC design to make the FPGA prototyping much easier. For a more in-depth discussion regarding the use of gated clocks for power optimizations, see Chapter 3.

6.2.1   Clocks Module

If a number of clocks are to be gated in an ASIC, it is recommended that all of these gating operations are consolidated into a single module dedicated to clock generation. This is illustrated in Figure 6.16.

Keep all gated clocks inside a dedicated clocks module and separate from the functional modules.

By keeping all the clock gating inside a single module, it makes the constraints easier to deal with as well as any modifications that must be made on the FPGA prototype. For instance, if the designer chooses to add a compile time macro to remove all gating elements for the FPGA, this can be easily organized within a single module. This is described in the following section.

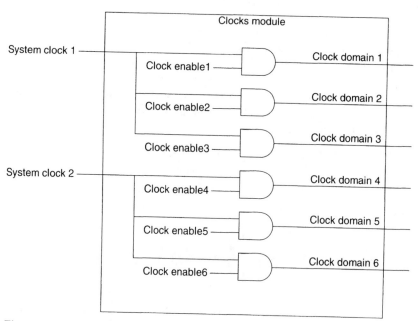

Figure 6.16   Consolidated clocks module.

6.2.2   Gating Removal

There are a number of slick ways to remove the clock gating for an FPGA prototype. The following example shows the most obvious, but cumbersome, method. This code removes all gating functions for the FPGA prototype.

```
`define FPGA
// `define ASIC

module clocks_block(...)

`ifdef ASIC
assign clock_domain_1 = system_clock_1 & clock_enable_1;
`else
assign clock_domain_1 = system_clock_1;
`endif
```

If the above format were used for the clocks module, only the define macro would need to be changed for the FPGA prototype. The downside is that a modification is required whenever targeting the FPGA prototype versus the ASIC. Many designers feel uncomfortable with this because they are not emulating the same RTL. A superior method would be to use an automatic gating removal tool to eliminate the probability of human error. Many modern synthesis tools will do this now with the proper constraints. For example, Synplify has an option called "fix gated clocks" which automatically moves the gating operation off of the clock line and into the data path. Consider the following example.

```
module clockstest(
    output reg oDat,
    input      iClk, iEnable,
    input      iDat);

    wire       gated_clock = iClk & iEnable;

    always @(posedge gated_clock)
        oDat <= iDat;
endmodule
```

In the above example, the system clock is gated with an enable signal to generate a gated clock. This gated clock is used to drive the flip-flop oDat, which registers the input iDat. Without fixing the clock gating, the synthesis tool will implement this directly.

In the implementation shown in Figure 6.17, the gating operation is placed on the clock line. Two clock domains now exist, must be constrained independently, and must be located on independent clocking resources. By enabling the clock gating removal, however, this gate is easily moved to the data path as shown in Figure 6.18.

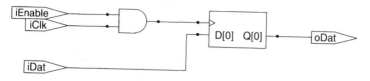

Figure 6.17 Direct clock gating.

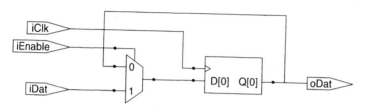

Figure 6.18 Clock-gating removal.

Most modern devices now provide a clock enable input which will eliminate the need for this type of solution. If a particular technology does not provide a flip-flop clock enable, however, this technique will add delay to the data path.

6.3 SUMMARY OF KEY POINTS

- Clock synchronization issues are generally not repeatable and will affect the reliability of the FPGA design.
- Metastability can cause catastrophic failures in the FPGA.
- The phase control technique can be used whenever the period of one clock is a multiple of the other and when one of the clocks can be controlled by an internal PLL or DLL.
- Double flopping can be used to resynchronize single-bit signals between two asynchronous clock domains.
- Timing analysis should ignore the first resynchronization flip-flop and ensure that the timing between the synchronization flip-flops themselves is minimized.
- FIFOs can be used when passing multibit signals between asynchronous clock domains.
- Gray codes can be used to pass multibit counter data between asynchronous clock domains and are often used inside FIFOs.
- Synchronization registers should be partitioned as independent blocks outside of the functional modules.
- If at all possible, avoid clock gating. If gating is necessary, keep all gated clocks inside a dedicated clocks module and separate from the functional modules.

Chapter 7

Example Design: I2S Versus SPDIF

The SPDIF (Sony/Philips Digital Interface Format) and I2S (Inter-IC Sound) standards have been developed and used by many consumer electronics manufacturers to provide a means for transmitting digital audio information between ICs and to eliminate the need to transmit analog signals between devices. By keeping the signal digital until the conversion to analog can be localized, it will be less susceptible to noise and signal degradation.

The objective of this chapter is to describe architectures for both I2S and SPDIF receivers and to analyze the method for recovery of the asynchronous signals and resynchronization of the audio data.

7.1 I2S

The I2S format is designed to transmit audio data up to sampling rates of 192 kHz in a source-synchronous fashion. By "source-synchronous" we are referring to the scenario where a clock is transmitted along with the data. With a source-synchronous signal, it is not necessary to share a system clock between the transmitting and receiving device. The sample size of the data can be 16 bits to 24 bits and is normalized to full-scale amplitude regardless of sample size. Unlike the SPDIF format, words of different lengths cannot be interchanged without defining the new size in the receiver.

The main design issue related to I2S is passing the samples between the source clock domain to the local clock domain. Because the signal is transmitted along with the source clock, the data can be easily reconstructed using the source clock and subsequently resynchronized.

Advanced FPGA Design. By Steve Kilts
Copyright © 2007 John Wiley & Sons, Inc.

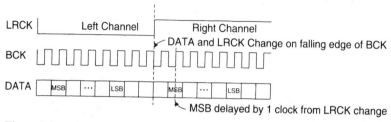

Figure 7.1   I2S timing.

7.1.1   Protocol

I2S has a very simple three-wire synchronous protocol. The three signals are defined as follows:

- LRCK (left/right channel select): When LRCK is low, the data belongs to the left channel, and when LRCK is high, the data belongs to the right channel.
- BCK (bit clock): This is the source-synchronous clock.
- DATA (serial audio data): This provides raw sample bits from the audio codes. The bits are synchronous with BCK.

The timing is illustrated with the waveforms shown in Figure 7.1.

As can be seen from these waveforms, LRCK defines the channel (low = left, high = right), and BCK clocks in the logic value on the DATA line. All transitions of the LRCK and DATA take place on the falling edge of the clock, which allows for a small amount of skew in either direction without violating setup and hold times. The length from the MSB to the LSB is defined by the word size, which is predefined in some manner depending on the application. Note that many I2S receivers have multiple modes outside of the "true" I2S format that are also considered a part of the protocol. These other formats include right and left justification mode, but here we will only consider the I2S format described above. Additionally, we will fix the data word size to 16 bits.

7.1.2   Hardware Architecture

The hardware architecture for an I2S module is very simple as shown in Figure 7.2.

On every rising edge of BCK, the logic value on DATA is clocked into the shift register. When a transition on LRCK is detected, the data word in the shift register is loaded into an output register determined by the polarity of LRCK. The entire I2S circuit uses BCK as the system clock to create a fully synchronous receiver. The data, once latched in the output register, must be passed to the local

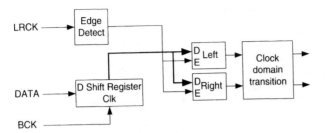

LRCK

DATA

BCK

Edge
Detect

D Shift Register
Clk

D Left
E

D Right
E

Clock
domain
transition

Figure 7.2 I2S architecture.

system clock domain. Thus, the domain transition occurs at the very end of the I2S data recovery. The implementation is shown below.

```
module I2S(
    output reg          oStrobeL, oStrobeR,
    output reg  [23:0]  oDataL, oDataR,
    input               iBCK, // bit clock
    input               iSysClk, // local system clock
    input               iDataIn,
    input               iLRCK);
    reg                 DataCapture;
    reg                 rdatain;
    // registers to capture input data on rising and falling
    // edges of clock
    reg         [23:0]  Capture;
    // strobes for valid data
    reg                 StrobeL, StrobeR;
    reg         [2:0]   StrobeDelayL, StrobeDelayR;
    reg         [23:0]  DataL, DataR;
    reg                 LRCKPrev;
    reg         [4:0]   bitcounter;
    reg                 triggerleft, triggerright;

    wire                LRCKRise, LRCKFall;
    wire        [23:0]  DataMux;

    // detect edges of LRCK
    assign LRCKRise = iLRCK & !LRCKPrev;
    assign LRCKFall = !iLRCK & LRCKPrev;

    // assuming 16 bit data
    assign DataMux  = {Capture[15:0], 8'b0};

    always @(posedge iBCK) begin
        DataCapture     <= (bitcounter != 0);
        triggerleft     <= LRCKRise;
        triggerright    <= LRCKFall;
        rdatain         <= iDataIn;
        // for detecting edges of LRCK
        LRCKPrev        <= iLRCK;
```

```verilog
      // capture data on rising edge, MSB first
      if(DataCapture)
        Capture[23:0]  <= {Capture[22:0], rdatain};

      // counter for left justified formats
      if(LRCKRise || LRCKFall)
        bitcounter    <= 16;
      else if(bitcounter != 0)
        bitcounter      <= bitcounter - 1;

      // Load data into register for resynchronization
      if(triggerleft) begin
        DataL[23:0]    <= DataMux;
        StrobeL        <= 1;
      end
      else if(triggerright) begin
        DataR[23:0]    <= DataMux;
        StrobeR        <= 1;
      end
      else begin
        StrobeL        <= 0;
        StrobeR        <= 0;
      end
    end


    // resynchronize to new clock domain
    always @(posedge iSysClk) begin
      // delay strobes relative to data
      StrobeDelayL    <= {StrobeDelayL[1:0], StrobeL};
      StrobeDelayR    <= {StrobeDelayR[1:0], StrobeR};

      // upon the rising edge of the delayed strobe
      // the data has settled
      if(StrobeDelayL[1] & !StrobeDelayL[2]) begin
        oDataL        <= DataL; // load output
        oStrobeL      <= 1; // single cycle strobe in
                            new domain
      end
      else
        oStrobeL      <= 0;

      if(StrobeDelayR[1] & !StrobeDelayR[2]) begin
        oDataR        <= DataR; // load output
        oStrobeR      <= 1; // single cycle strobe in new
                            domain
      end
      else
        oStrobeR      <= 0;
    end
endmodule
```

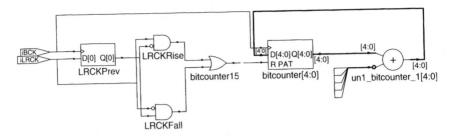

Figure 7.3 LRCK detection.

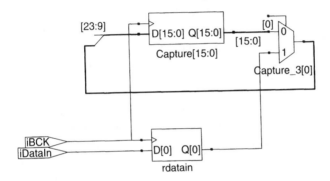

Figure 7.4 Bit capture.

The first step in the above implementation is to detect a transition on LRCK so we can clear the bit counter. This is implemented in a synchronous fashion as shown in Figure 7.3.

Next, we need to begin capturing bits into our shift register as shown in Figure 7.4.

Finally, we use the LRCK trigger to load the shift register into the output register and resynchronize the data with the local clock domain.

7.1.3 Analysis

When capturing and resynchronizing data from a source-synchronous data stream, there are a number of options available to the designer. The three options available with the I2S implementation are

1. Using a delayed validity bit to resynchronize the outputs

2. Double flopping the input stream

3. FIFO outputs

In the above implementation, we chose to use a delayed validity bit. Note that there are a number of design considerations when choosing a method for a particular implementation. The first consideration is speed. The advantage of the above implementation is that it runs at the audio bit clock speed, which in the worst case (192 kHz) is about 12 MHz. If we were running this module at the system clock speed, we may have to meet timing at perhaps hundreds of megahertz. Clearly, timing compliance will be much easier at the slower clock speed, which will allow the designer flexibility to implement low-area design techniques and allow the synthesis tool to target a compact implementation. The disadvantage is the increased complexity of the clock distribution and timing analysis. The implementation results are shown for each topology at the end of this section.

The scenario where a FIFO would be required at the outputs would arise when the receiving system (located behind the I2S interface) cannot handle periodic bursts of data. If the hardware were a pure pipeline or was at least dedicated to the processing of the incoming audio data, this would not be a problem. However, if the device that is capturing the data accesses the module through a shared bus, the data cannot simply present itself as soon as it is available. In this case, a FIFO provides a clean transition to the new domain as long as the average data rate on the bus end is greater than the audio data rate as shown in Figure 7.5.

The implementation of Figure 7.5 will require dual-port RAM resources as well as some control logic to implement the FIFOs. The final implementation results for all topologies are shown in Table 7.1.

Clearly, there is a significant amount of overhead associated with the FIFO implementation and it would not be a desirable solution unless required by the system.

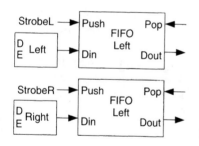

Figure 7.5   FIFO synchronization.

Table 7.1   Implementation Results for I2S Synchronization

Frequency	Double-flop outputs 197 MHz	Double-flop inputs 220 MHz	FIFO outputs 164 MHz
Flip-flops	62	72	130
LUTs	15	35	62
Clock buffers	2	1	2
Block RAMs	0	0	2

7.2 SPDIF

The SPDIF format is designed to transmit audio data up to sampling rates of 192 kHz (until recently, the maximum sampling frequency has been locked at 96 kHz, so many devices will not upsample beyond this prior to transmission). The sample size of the data can be 16 bits to 24 bits and is normalized to full-scale amplitude regardless of sample size. In other words, additional bits are automatically detected as additional bits of precision and not an increase in absolute amplitude. From an implementation perspective, a 16-bit word can be viewed as a 24-bit word with 8 bits of zeros appended to the least significant bits of precision. Thus, capturing the data word is the same regardless of word size (contrast this with I2S, which must have word size and format defined prior to capture).

The main design issue related to SPDIF is its asynchronous nature. Because the signal is transmitted via only one wire, there is no way to directly synchronize to the transmitting device and ultimately the audio signal. All of the information necessary to recover the clock is encoded into the serial stream and must be reconstructed before audio information can be extracted.

7.2.1 Protocol

Each sample of audio data is packetized into a 32-bit frame that includes additional information such as parity, validity, and user-definable bits (the user bits and even the validity bits are often ignored in many general-purpose devices). For stereo applications, two frames must be transmitted for each sample period. Thus, the bit rate must be $32*2*F_s$ (2.8224 MHz for 44.1 kHz, 6.144 MHz for 96 kHz, etc). The 32-bit packet format is defined in Table 7.2.

In the implementation described in this chapter, we will only decode the audio data and preamble.

To enable the SPDIF receiver to identify distinct bits as well as to resynchronize the packets, a special one-wire encoding is used called Biphase Mark Code (BMC). With this form of encoding, the data signal transitions on every bit regardless of whether it is encoded as a 1 or a 0. The difference between these

Table 7.2   SPDIF Frame Definition

Bits	Field
31	Parity (not including the preamble)
30	Channel status information
29	Subcode data
28	Validity (0 = valid)
27 : 4	Audio sample (MSB at bit 27)
3 : 0	Preamble

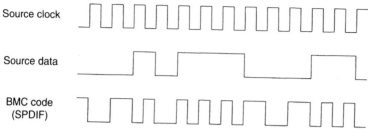

Source clock

Source data

BMC code
(SPDIF)

Figure 7.6   Example BMC encoding.

Table 7.3   SPDIF preambles

Preamble	SPDIF signal if last level = 0	SPDIF signal if last level = 1
Left channel at the start of a data block	11101000	00010111
Left channel not at the start of a data block	11100010	00011101
Right channel	11100100	00011011

bits is that the SPDIF signal will transition once per bit for a logic-0 and twice per bit for a logic-1. An example encoding is shown in Figure 7.6.

The first two waveforms shown in Figure 7.6 are the clock and data seen by the transmitter. In a synchronous transmission medium such as I2S, this clock as well as the synchronized data are passed to the receiver making the data recovery trivial. When only one wire is available, the data is encoded in BMC format as shown in the third waveform. As can be seen from this waveform, the clock is encoded into the data stream with the requirement of at least one transition for every bit. Note that the clock that sources the SPDIF stream must be twice the frequency of the audio clock to provide two transitions for every logic-1.

Due to the fact that the encoding of a data bit must transition once per bit, SPDIF provides a means to synchronize each frame by violating this condition once per frame. This is performed in the preamble as shown in Table 7.3.

As can be seen from these bit sequences, each preamble violates the transition rule by allowing a sequence of three consecutive clock periods of the same level. Detecting these preambles allows the receiver to synchronize the audio data to the appropriate channel. For a hardware implementation, a clock with a sufficient frequency must be used to be able to not only distinguish the difference between a logic-0 and a logic-1 (a 2× difference in pulse widths) but also a difference between a logic-0 and a preamble (a 1.5× difference in pulse widths).

7.2.2   Hardware Architecture

The basic architecture for the SPDIF receiver is shown in Figure 7.7.

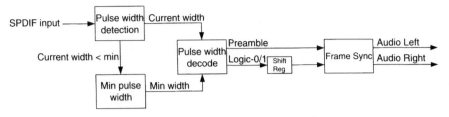

Figure 7.7   SPDIF architecture.

The pulse width detection logic contains a free-running counter that resets whenever the input from the BMC code toggles. In addition to the counter reset, the current width is decoded against the running minimum pulse width. If the current width is greater than $2.5\times$ the minimum pulse width, the pulse is decoded as a BMC violation and part of the preamble. If the width is greater than $1.5\times$ the min pulse width, the pulse is decoded as a logic-0. If the width is less than the running minimum pulse width, it overrides the minimum pulse width, and the audio data is assumed to be invalid due to the absence of a lock. Otherwise, the pulse is decoded as half of a logic-1.

If a logic-1 or logic-0 is detected, this bit is shifted into a 24-bit shift register in preparation for synchronization with the preamble. When a preamble is detected, the previous frame has completed and can now be decoded based on the mapping of the various fields. The implementation is shown in the following code.

```
module spdif(
    output reg          oDatavalidL, oDatavalidR,
    output reg [23:0]   oDataL, oDataR,
    input               iClk, // main system clock used to
                             sample spdif data
    input               iSPDIFin);
    reg       [2:0]     inputsr; // input shift register
    reg                 datatoggle; // register pulses high
                             when data toggles
    // counts the width between data transitions
    reg       [9:0]     pulsewidthcnt;
    // register to hold width between transitions
    reg       [9:0]     pulsewidth;
    reg       [9:0]     onebitwidth; // 1-bit width reference
    // signals that pulsewidth has just become valid
    reg                 pulsewidthvalid;
    reg                 bitonedet; // detect logic-1 capture
    reg                 newbitreg; // new data registered
    reg       [27:0]    framecapture; // captured frame
    reg                 preambledetect;
    reg                 preamblesyncen;
    reg                 channelsel;  // select channel based
                             on preamble
```

```
reg          [5:0]  bitnum;
reg          [10:0] onebitwidth1p5;

reg                 onebitload; // load 1-bit reference
                                width
reg                 onebitupdown; // 1: reference width
                                  should increment
// width used for comparison against reference
reg          [9:0]  pulsewidthcomp;
reg                 onebitgood; // reference is equal to
                                input width
reg                 preamblesync; // flags preamble in
                                  spdif stream
reg                 shiftnewdat; // ok to capture
// load data into output buffer
reg                 outputload, outputloadprev;
reg                 pulsewidthsmall, pulsewidthlarge;
reg          [11:0] onebitwidth2p5;
wire                trigviolation;
wire                newbit;   // raw data decoded from stream

// flag a violation in BMC code
assign trigviolation   = {1'b0, pulsewidth[9:0], 1'b0} >
                          onebitwidth2p5;

// if width is small, data is 1.  Otherwise data is 0
assign newbit          = ({pulsewidth[9:0],1'b0} <
                          onebitwidth1p5[10:0]);

always @(posedge iClk) begin
  inputsr              <= {inputsr[1:0], iSPDIFin};
// shift data in
// trigger on change in data
  datatoggle           <= inputsr[2] ^ inputsr[1];

// counter for pulse width
  if(datatoggle) begin
// counter resets when input toggles
    pulsewidth[9:0]    <= pulsewidthcnt[9:0];
    pulsewidthcnt      <= 2;
  end
  else
    pulsewidthcnt      <= pulsewidthcnt + 2;

  // width register will be valid 1 clock after the data
    toggles
  pulsewidthvalid      <= datatoggle;

  // onebitload checks to see if input period is out of
    bounds
  // current width is 1/2 1-bit width
  pulsewidthsmall      <= ({1'b0, onebitwidth[9:1]} >
                          pulsewidth[9:0]);
```

```
// current width is 4x 1-bit width
pulsewidthlarge      <= ({2'b0, pulsewidth[9:2]} >
                         onebitwidth);
// load new reference if out of bounds
onebitload           <= pulsewidthlarge || pulse
                         widthsmall;

// register width comparison value
if(!newbit)
   pulsewidthcomp    <= {1'b0, pulsewidth[9:1]};
else
   pulsewidthcomp    <= pulsewidth[9:0];

// checks to see if reference is equal to input width
onebitgood           <= (pulsewidthcomp == onebit
                         width);
// increment reference if input width is greater than
   reference
onebitupdown         <= (pulsewidthcomp > onebitwidth);

// keep track of 1-bit width
// load reference if input width is out of bounds
if(onebitload)
   onebitwidth       <= pulsewidth[9:0];
else if(!onebitgood && pulsewidthvalid) begin
   // adjust reference
   if(onebitupdown)
      onebitwidth     <= onebitwidth+1;
   else
      onebitwidth     <= onebitwidth-1;
end
// set onebitwidth*1.5 and onebitwidth*2.5
onebitwidth1p5       <= ({onebitwidth[9:0], 1'b0} +
{1'b0, onebitwidth[9:0]});
onebitwidth2p5       <= ({onebitwidth[9:0], 2'b0} +
{2'b0, onebitwidth[9:0]});
// preamblesync is valid only when last frame has
   completed
preamblesyncen       <= (bitnum == 0) && datatoggle;
// trigger on preamble in spdif header if input width
   > 2.5*reference
preamblesync         <= preamblesyncen && trigviolation;

// capture preamble
if(preamblesync)
   preambledetect    <= 1;
else if(preambledetect && pulsewidthvalid)
   preambledetect    <= 0;

// set channel
if(preambledetect && pulsewidthvalid)
```

```
        channelsel          <= !trigviolation;
      else if(trigviolation && pulsewidthvalid)
        channelsel          <= 0;

      newbitreg             <= newbit;
      // only trigger on a bit-1 capture every other transition
      if(!newbitreg)
        bitonedet           <= 0;
      else if(newbit && datatoggle)
        bitonedet           <= !bitonedet;

      // set flag to capture data when bit-0 or bit-1 is valid
      shiftnewdat           <= pulsewidthvalid && (!newbit ||
                                 bitonedet);

      // shift register for capture data
      if(shiftnewdat)
        framecapture[27:0] <= {newbit, framecapture[27:1]};

      // increment bit counter when new bit is valid
      // reset bit counter when previous frame has finished
      if(outputload)
        bitnum              <= 0;
      else if(preamblesync)
        bitnum              <= 1;
      else if(shiftnewdat && (bitnum != 0))
        bitnum              <= bitnum + 1;

      // data for current frame is ready
      outputload            <= (bitnum == 31);
      outputloadprev        <= outputload;

      // load captured data into output register
      if(outputload & !outputloadprev) begin
        if(channelsel) begin
          oDataR            <= framecapture[23:0];
          oDatavalidR       <= 1;
        end
        else begin
          oDataL            <= framecapture[23:0];
          oDatavalidL       <= 1;
        end
      end
      else begin
        oDatavalidR         <= 0;
        oDatavalidL         <= 0;
      end
    end
  end
endmodule
```

The first step in the above architecture is to resynchronize the incoming data stream to the local system clock. A double-flop technique is used as described in previous chapters for passing a single bit across domains. This is shown in Figure 7.8.

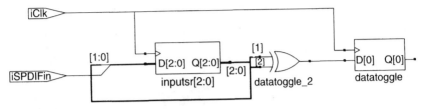

Figure 7.8   Resynchronizing the SPDIF input.

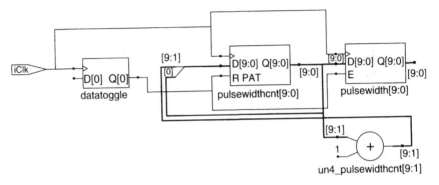

Figure 7.9   SPDIF pulse width counter.

Note that the bits that are used for edge detection are bits 2 and 1. Bits 0 and 1 in the shift register are used for clock synchronization only, and bit 2 is used for the detection of a transition. The synchronized toggle flag in datatoggle is used to reset the counter for the pulse width as shown in Figure 7.9. Notice how the synthesis tool was able to utilize the reset and clock enable pins of the flip-flop elements and eliminate any muxing. This was described in Chapter 2.

The next step is to determine if the pulse width is out of acceptable bounds and whether we need to reset the running value for a 1-bit width. The logic shown in Figure 7.10 performs the boundary condition checks and sets a bit to reload the reference width.

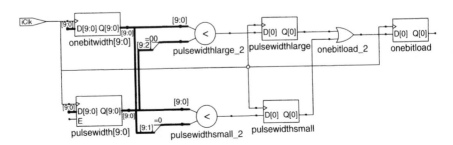

Figure 7.10   Pulse width reference.

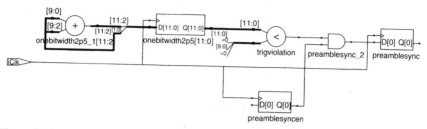

Figure 7.11   Preamble detection.

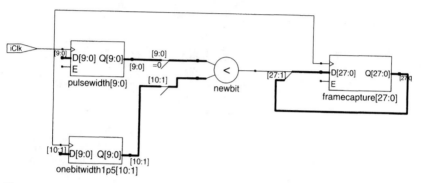

Figure 7.12   Bit detection.

The next block of logic is to detect a preamble. Figure 7.11 shows the scaling of the reference width by 2.5 and performing of the frame synchronization.

Note in the implementation of Figure 7.11 that the factor of 2.5 was optimally implemented by a simple shift and add of the original signal. Similarly, we need to determine if the pulse width is indicating a bit-0 or a bit-1 (assuming the pulse width is not indicating a preamble).

In the circuit shown in Figure 7.12, the data that is shifted into the frame-capture shift register is dependent on the width of the current pulse. In other words, if the current pulse width is less than 1.5× the pulse width of the value of a single bit width, the data shifted in is a logic-1. Otherwise, the data is a logic-0.

Finally, a transition on the output load is detected (dependent on the bit counter), the channel is selected, and the frame data is loaded into the appropriate output register as shown in Figure 7.13.

7.2.3 Analysis

When resynchronizing a signal with an encoding such as BMC, there is no choice but to sample this signal at the front end and map it into the local clock domain. No processing can take place until this initial resynchronization occurs. Additionally, the system clock that is used to sample the SPDIF stream must be sufficiently faster than the minimum pulse width of the SPDIF stream itself to provide

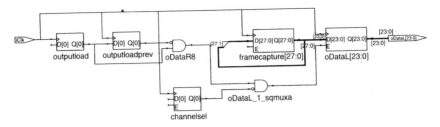

Figure 7.13 SPDIF output Synchronization

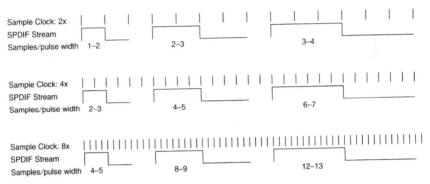

Figure 7.14 SPDIF sampling rates.

enough resolution when detecting thresholds in the pulse width. Specifically, under all relative phases of the sampling clock to the SPDIF stream, we require the following:

- The pulse width of a logic-0 is between $1.5\times$ and $3\times$ of the minimum pulse width (logic-1).
- The pulse width of a preamble violation is between $2.5\times$ and $4\times$ of the minimum pulse width.
- There are at least two clock periods of margin in the thresholds to account for jitter of either the input stream or the system clock.

Figure 7.14 illustrates the various sampling rates.

As can be seen from this diagram, the criteria for reliable signal recovery is when we have a sampling rate of at least $8\times$ the maximum clock frequency (full

Table 7.4 Implementation Results in a Xilinx Spartan-3 XC3S50

Frequency	130 MHz
FFs	161
LUTs	153

period for a logic-1). For a 192-kHz sampling rate, this corresponds with a worst-case timing of: 192 kHz-64*8 = 98.304 MHz. If we target this at a Xilinx Spartan-3 device with a 10 ns period (allowing for about 100 ps of jitter), we obtain the results shown in Table 7.4.

Although we can easily achieve the desired frequency, the logic required to implement the signal recovery is large relative to a source-synchronous system such as I2S.

Chapter 8

Implementing Math Functions

This chapter covers a variety of problems encountered when an FPGA designer attempts to implement a complex math function in an FPGA. Interestingly, most real-world math problems can be solved by combinations of shift and add operations. This chapter describes a few of these methods as they relate to division and trigonometric operations and then also explains how to expand this to a broader class of functions. In many cases, there are a number of alternative solutions that require optimizations for a given application.

During the course of this chapter, we will discuss the following topics:

- Methods for performing efficient division for both fixed and floating point operations.

  Basic multiply and shift algorithm for fixed divisors
  Iterative division for fixed point operations
  The Goldschmidt method for high-throughput, pipelined division operations
- Successive approximation using Taylor and Maclaurin series.
- CORDIC algorithm for trigonometric functions.

8.1 HARDWARE DIVISION

Division operations have always caused headaches for digital logic designers. Unlike addition, subtraction, or multiplication, there is no simple logic operation that will generate a quotient. Among other subtle difficulties, division differs from the other arithmetic operations in that fixed-point operations do not produce a finite and predictable fixed-point result. There are, however, a number of ways to deal with this issue. The simplicity of the solution will be entirely application dependent, but for the sake of discussion we will begin with the simple solutions requiring specific constraints and then move on to general solutions.

Advanced FPGA Design. By Steve Kilts
Copyright © 2007 John Wiley & Sons, Inc.

8.1.1 Multiply and Shift

The multiply and shift method is the simplest solution to the division problem and is essentially equivalent to multiplying by the inverse of the divisor.

This solution leverages the basic property of binary numbers that a shift to the least-significant position (a right shift in most HDL representations) will result in a divide-by-two. If an 8-bit register has a fixed point at bit 4 and contains the integer value of 3, the register will have the representation shown in Figure 8.1.

With the fixed-point representation defined earlier, the number represented in Figure 8.1 is equivalent to: $2^1 + 2^0 = 3$. If we divide by 2, we can perform a right shift by 1 as shown in Figure 8.2.

The register of Figure 8.2 now holds the value $2^0 + 2^{-1} = 1.5$. A simple multiply and shift technique assumes the divisor is fixed to some constant and defines the divide operation as a multiply followed by a divide by some power of 2. For instance, a divide by 7 could be approximated by a multiplication by 73 followed by a divide by 512 (implemented with a 9-bit right shift). The result is a divide by $7.013\ldots$. Greater precision can be achieved by increasing the power of 2 and increasing the multiplication factor appropriately.

As mentioned previously, this is essentially equivalent to inverting the divisor and multiplying. Specifically, the divisor is inverted and converted to a unique fixed-point number (fixed point at the ninth bit in the above example) and then converted back to the original fixed-point format. This technique works well for high-speed divisions by constant factors or for applications where only a few bits of precision are required (many bits of precision require large multiply operations).

> The multiply and shift method is an easy way to perform division but can only be used when the divisor is represented in a specific form.

It should be noted that to represent the fractional number properly, some other device such as an external microprocessor will need to provide an initial

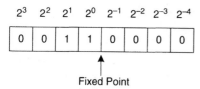

Figure 8.1 Fixed-point representation of 3.

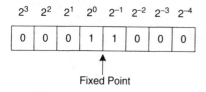

Figure 8.2 Fixed-point representation of 1.5.

inversion. If an external device is defining the divisor, this method can work well. If the divisor is determined by some other logic and the design does not have the luxury of this representation, then other methods will be required as described in the following sections.

8.1.2 Iterative Division

The division algorithm discussed in this section is an example of an algorithm that falls into a class of digit recurrence methods. Iterative methods typically refer to successive approximation methods (discussed in the next section), but for consistency with other chapters, we will refer to this as an iterative method as it maps to similar iterative methods used for compact implementations of other functions.

Iterative division works much like long division with decimal numbers did in elementary school. One of the advantages in using binary numbers, however, comes from the fact that certain optimizations can be made in the division process. Take the following long division example with the binary representations of 2 divided by 3 as shown in Figure 8.3.

The divider for a fixed-point number can be architected as shown in Figure 8.4 with a comparator and a subtraction unit. With this architecture, the dividend is "normalized" to a fixed-point value that is necessarily smaller than twice the divisor. By doing this, every subsequent shift operation will produce a new partial quotient that must be less than twice the divisor. This means that the divisor will "go into" the partial quotient 1 or 0 times. If the divisor is less than or equal to the partial dividend for the current iteration, a logic-1 is shifted into the quotient register, otherwise, a logic-0 is shifted in and the partial dividend is shifted left by 1. After the necessary number of iterations to achieve the desired precision, the output is postnormalized to shift the data to the proper fixed-point location (undo the prenormalization).

```
       0.1010101...
11 | 10
     00
     100
      11
       1...
```

Figure 8.3 Long division in binary.

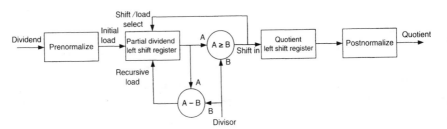

Figure 8.4 Simple-fixed point division architecture.

Because of the frequency with which this arises, advanced synthesis tools such as Synplify Pro will automatically implement this type of structure for fixed-point operations. If an integer is declared, Synplify Pro will use a 32-bit word and automatically optimize unused bits.

> This compact architecture is useful for fixed-point divisions that can afford the relatively large latency of an iterative division process.

If more space can be afforded and faster divide operations are required between arbitrary numbers, more sophisticated techniques are required as described in the following section.

8.1.3 The Goldschmidt Method

The Goldschmidt method can be used when high-speed division operations need to be pipelined to maximize throughput. This method falls into a class of successive approximation algorithms (sometimes called iterative, but not to be confused with the implementation from the previous section) that approach the true quotient with each recursion of the algorithm. The nice feature about algorithms such as Goldschmidt is that they can be pipelined in an efficient manner. In other words, the hardware can be constructed such that one division operation can be completed on each edge of the clock and can be implemented with less area than would be required for a brute-force unrolling of the iterative method discussed in the previous section.

> The Goldschmidt method provides a way to pipeline the division process in a manner that is much more efficient than unrolling the loop of the iterative method.

The idea behind the Goldschmidt algorithm is to calculate $Q = N/D$ by approximating the value $1/D$ for multiplication with N, and then approach the true quotient through successive approximations. This is useful for large numbers such as those represented by the IEEE 754 floating-point standard, which has 64-bit operators. Clearly, a look-up table of the order 2^{50} bits (potential size of a floating-point mantissa) is infeasible, but drastically reduced look-up tables of the order 2^{10} (a typical value for a Goldschmidt implementation) can be practical.

There are a number of papers that provide a fine theoretical foundation for the Goldschmidt algorithm. Of particular interest to the design engineer is the number of iterations required to produce an error within acceptable bounds. The steps to approximate $Q = N/D$ using the Goldschmidt algorithm are as follows:

1. Move the fixed point for N and D to locations such that $N \geq 1$ and $D < 2$. In the context of floating-point operations, this is referred to as normalizing the numerator and denominator.

2. Start with an initial approximation to $1/D$ by use of a look-up table and call it L_1. Eight to 16 bits of precision is often enough depending on the application.

3. Calculate the first approximation to $q_1 = L_1 N$ and the error term $e_1 = L_1 D$ (which will approach 1 as the iterations approach infinity).

4. Iteration begins by assigning $L_2 = -e_1$ (two's complement).

5. $e_2 = e_1 L_2$ and $q_2 = q_1 L_2$.

6. $L_3 = -e_2$ similar to step 4 and continue with successive iterations.

After each iteration of this algorithm, e_i approaches 1 (the denominator D multiplied by 1/D), and q_i approaches the true quotient Q. To calculate the bounded error value based on the number of iterations and number of bits in your system, refer to one of the many papers that discusses this algorithm from a more rigorous sense. In practice, 4–5 iterations often provides sufficient accuracy for 64-bit floating-point (53-bit fixed-point) calculations.

The following example illustrates the use of the Goldschmidt algorithm.

```
module div53(
    output [105:0]  o, // quotient
    input           clk,
    input  [52:0]   a, b); // dividend and divisor
    reg    [261:0]  mq5;
    reg    [65:0]   k2;
    reg    [130:0]  k3;
    reg    [130:0]  k4;
    reg    [130:0]  k5;
    reg    [52:0]   areg, breg;
    reg    [65:0]   r1, q1;
    reg    [130:0]  r2, q2;
    reg    [130:0]  r3, q3;
    reg    [130:0]  q4;
    wire   [13:0]   LutOut;
    wire   [13:0]   k1;
    wire   [66:0]   mr1, mq1;
    wire   [131:0]  mr2, mq2;
    wire   [261:0]  mr3, mq3;
    wire   [261:0]  mr4, mq4;

    gslut gslut(.addr(b[51:39]),
                .clk(clk),
                .dout(LutOut));

    assign k1   = LutOut;

    assign o    = mq5[261-1:261-1-105];

    assign mr1 = breg * k1;
    assign mq1 = areg * k1;

    assign mr2 = r1   * k2;
    assign mq2 = q1   * k2;

    assign mr3 = k3   * r2;
    assign mq3 = k3   * q2;

    assign mr4 = k4   * r3;
    assign mq4 = k4   * q3;
```

```
always @(posedge clk) begin
  areg  <= a;
  breg  <= b;

  r1    <= mr1[65:0];
  k2    <= ~mr1[65:0] + 1;
  q1    <= mq1[65:0];

  r2    <= mr2[130:0];
  k3    <= ~mr2[130:0] + 1;
  q2    <= mq2[130:0];

  r3    <= mr3[260:130];
  k4    <= ~mr3[260:130] + 1;
  q3    <= mq3[260:130];

  k5    <= ~mr4[260:130] + 1;
  q4    <= mq4[260:130];

  mq5   <= k5 * q4;
end
endmodule
```

In the above example, a 53-bit division is performed (as may be required by the IEEE 754 standard for 64-bit floating-point numbers) using a fully pipelined architecture. The inputs are assumed to be normalized (and thus instantiated somewhere in the hierarchy), and no overflow checks are performed. It is noted that this is fully expanded for maximum speed (one operation per clock) but will also be relatively large when synthesized. Possible area optimizations include the use of a single multiplier using a state machine to iterate through the product coefficients and/or the use of a compact multiplier that uses repetitive shift-add operations.

8.2 TAYLOR AND MACLAURIN SERIES EXPANSION

Taylor and Maclaurin series expansions can be used to break down operations such as exponentials, trigonometric functions, and logarithms into simple multiply and add operations that are better suited for hardware. The general form of the Taylor expansion is shown in Equation (8.1),

$$T(x) = \sum_{0}^{\infty} \frac{f^{(n)}(a)(x-a)^n}{n!} \qquad (8.1)$$

where $f^{(n)}$ is the n-th derivative of f. Commonly in practice, $a = 0$, and the above expansion simplifies to the Maclaurin series as shown in Equation (8.2):

$$M(x) = \sum_{0}^{\infty} \frac{f^{(n)}(0)(x)^n}{n!} \qquad (8.2)$$

$$\sin(x) = x - \frac{x^3}{3!} + \frac{x^5}{5!} - \cdots$$

$$\cos(x) = 1 - \frac{x^2}{2!} + \frac{x^4}{4!} - \cdots$$

$$\ln(1+x) = x - \frac{x^2}{2} + \frac{x^3}{3} - \cdots$$

$$e^x = 1 + x + \frac{x^2}{2!} + \frac{x^3}{3!} + \cdots$$

Figure 8.5   Useful expansions.

Methods for creating the expansion functions are covered in many other texts, and in practice the most common functions are already well defined. Some useful expansions are listed in Figure 8.5.

Figure 8.6 illustrates the sine wave approximation as the order of the series expansion is increased. Clearly, the accuracy as well as the desired range will determine the order of the approximation.

From the expansions of Figure 8.5, the usefulness of Taylor and Maclaurin series expansions should be readily evident. All denominators are fixed numbers

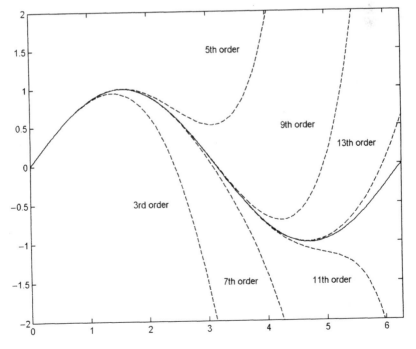

Figure 8.6   Sine wave approximations.

that can be inverted ahead of time and applied as a fixed-point multiplication as described in previous sections.

Taylor and Maclaurin series expansions can be used to break down complex functions into multiply and add operations that are easily implemented in hardware.

The primary drawbacks to a series expansion algorithm are both the number of multiplies required and the time required to properly iterate. These two are often related as a very compact architecture would require shift-and-add multipliers that set the algorithm into potentially hundreds of clock cycles. The next section describes an algorithm that uses binary approximation via vector rotation that can dramatically increase the speed of the approximation.

8.3 THE CORDIC ALGORITHM

The CORDIC (coordinate rotation digital computer) method is a successive approximation algorithm that is useful for calculating the trigonometric functions sine and cosine very efficiently. CORDIC uses a sequence of vector rotations to calculate trigonometric functions with successive approximation. To conceptualize, consider the following graph-based technique for calculating sine and cosine:

1. Draw a vector on the x–y plane with a magnitude of 1 and a phase of 0 as shown in Figure 8.7.

2. Begin rotating the vector counterclockwise until the desired angle is reached. Maintain a magnitude of 1 as shown in Figure 8.8.

3. Note the (x, y) coordinates at the desired angle. The sine is simply the y value, and the cosine is the x value (the hypotenuse is 1, so $\sin = y/1$ and $\cos = x/1$). This is illustrated in Figure 8.9.

In a hardware implementation, the vector rotation is performed by making adjustments in increments of 90 degrees divided by successively larger powers of 2 (successively smaller angle increments) and updating the x–y coordinates for each jump. The decision to add or subtract the current incremental value depends on where the algorithm is relative to the target angle. Thus, we successively

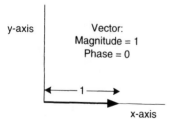

Figure 8.7 CORDIC initialization.

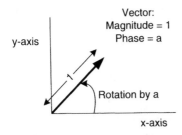

Figure 8.8   CORDIC rotation.

approach the desired angle in progressively smaller increments. The iterative equations are defined as follows:

$$x_{i+1} = K_i[X_i - y_i d_i 2^{-i}]$$

$$y_{i+1} = K_i[y_i + x_i d_i 2^{-i}]$$

$$K_i = (1 + 2^{-2i})^{-1/2}$$

Note that K_i is a fixed value determined by the iteration stage and approaches 0.60725... as i approaches infinity. The decision term d_i is 1 if the target angle is greater than the cumulative angle (increase the angle by increasing y and decreasing x). Similarly, d_i is -1 if the target angle is less than the cumulative angle (decrease the angle by decreasing y and increasing x). In a practical application, the designer would calculate K_i ahead of time based on the number of iterations and apply this at the end of the calculation as a constant factor. For a theoretical proof of these equations, the reader may refer to one of the many papers published on the topic of CORDIC theory.

For most implementations, the recursion operates on add/subtract and comparison operations, all of which can easily be performed in a single clock cycle. Thus, the algorithm will run much faster or for a given speed can be implemented with fewer gates than required by a Taylor approximation. The only multiply operations are those that occur once at the very end of the calculation and are constant factor multiplications that can be further optimized.

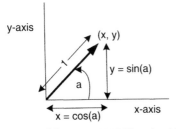

Figure 8.9   Final CORDIC angle with sine and cosine.

The CORDIC algorithm should be preferred over a Taylor expansion for the calculation of sine and cosine operations.

8.4 SUMMARY OF KEY POINTS

- The multiply and shift method is an easy way to perform division but can only be used when the divisor is represented in a specific form.
- This compact architecture is useful for fixed-point divisions that can afford the relatively large latency of an iterative division process.
- The Goldschmidt method provides a way to pipeline the division process in a manner that is much more efficient than unrolling the loop of the iterative method.
- Taylor and Maclaurin series expansions can be used to break down complex functions into multiply and add operations that are easily implemented in hardware.
- The CORDIC algorithm should be preferred over a Taylor expansion for the calculation of sine and cosine operations.

Chapter 9

Example Design: Floating-Point Unit

$\mathbf{M}$ost of the implementations in the previous chapter described manipulations of fixed-point numbers. If calculations are taking place over a very wide range of numbers, however, a floating-point representation may be required. The floating-point unit (FPU) described in this section is a hardware model for the IEEE 754-1985 floating-point standard. The objective of this chapter is to describe a pipelined architecture for implementing add and subtract operations and to analyze the implementation.

9.1 FLOATING-POINT FORMATS

According to the IEEE standard, a floating-point number contains a sign bit, an exponent, and a mantissa. For the 32-bit standard, the format is illustrated in Figure 9.1.

When not considering the boundary conditions or the maximization of the full numerical range of a given floating-point representation, floating-point add and subtract operations are very simple to implement via manipulation of the mantissa and exponents. However, for compliance with the IEEE standard, the aforementioned considerations must be taken into account. In particular, a region called the "subnormal region" is a provision in the standard to provide additional precision for numbers near the low end of the numerical representation. The normal and subnormal regions are defined in Table 9.1.

Note that there are a number of other of conditions including infinity, NaN (not-a-number), and so forth. We will not discuss these additional formats in this chapter.

Advanced FPGA Design. By Steve Kilts
Copyright © 2007 John Wiley & Sons, Inc.

Field	Sign	Exponent	Mantissa
Bits	31	30 23	22 0

Figure 9.1 32-bit floating point representation.

Table 9.1 Normal and Subnormal Regions

Region	Condition	Representation
Normal	$0 < \text{exponent} < 255$	$(-1)^s \times 2^{e-127} \times 1.m$
Subnormal	$\text{exponent} = 0$	$(-1)^s \times 2^{-126} \times 0.m$

9.2 PIPELINED ARCHITECTURE

The FPU is implemented with a fully pipelined architecture. This architecture maximizes performance for high-speed applications, allowing the user to apply new inputs on every clock edge. Figure 9.2 illustrates the various functional blocks and the flow of the pipeline.

The first step is to detect whether we are operating in the subnormal region (with exponent = 0). The circuit shown in Figure 9.3 appends a logic-1 to the MSB of the mantissa if we are in the normal region and a logic-0 if we are in the subnormal region.

Next, the smaller of the two numbers must be normalized such that the exponents of the mantissas are equal.

In Figure 9.4, the mantissa of the smaller number is shifted by the difference between the two exponents. The addition/subtraction operation can now operate on the two mantissas. If the sign of the two floating-point numbers are the same, the mantissas can add, but otherwise they will subtract. The logic is shown in Figure 9.5.

Finally, the resulting mantissa must be shifted into the proper floating-point format (1.xxx for normal representation, 0.xxx for subnormal). The mantissa is postnormalized, and the resulting shift is subtracted from the exponent. This is shown in Figure 9.6.

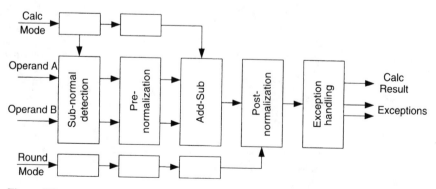

Figure 9.2 Fully pipelined FPU.

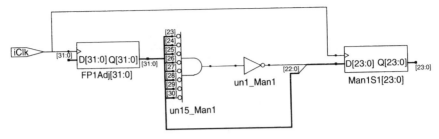

Figure 9.3 Subnormal detection.

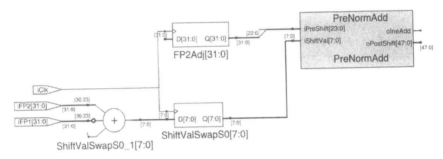

Figure 9.4 Mantissa normalization.

All overflow and special condition operations are performed with comparators and multiplexers as shown in the final Verilog implementation but are not shown in these logic diagrams. Finally, because the design is pipelined, a new floating-point operation may take place on every clock edge. The following list describes the steps necessary to perform a floating-point addition:

1. Present the operands OpA and OpB to the input. They will be clocked in on the next rising edge of Clk.

2. Present the control signals AddSub and Rmode (rounding mode) to the input synchronous to OpA and OpB. These four inputs define an operation that will begin on the next rising edge of Clk.

3. When the Add-Sub operation has completed, the operation output becomes valid synchronous with all of the condition flags.

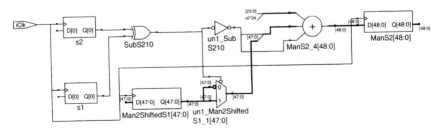

Figure 9.5 Sign detection.

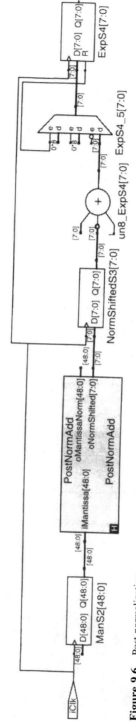

Figure 9.6 Post-normalization.

9.2.1 Verilog Implementation

```
// Floating point addition according to the IEEE 754 standard.
//

`define ROUNDEVEN    2'b00
`define ROUNDZERO    2'b01
`define ROUNDUP      2'b10
`define ROUNDDOWN    2'b11

// other
`define INF          31'b1111111110000000000000000000000
`define INDVAL       31'b1111111111000000000000000000000
`define LARGESTVAL   31'b1111111101111111111111111111111

module FPAdd(
    output [31:0] oFPSum,              // floating
                                       //   point sum

    output        oIneAdd,            // inexact
                                      //   addition

    output        oOverflowAdd,       // overflow
                                      //   from addition

    input         iClk,
    input  [31:0] iFP1, iFP2,         // floating
                                      //   point inputs

    input  [1:0]  iRMode,             // rounding
                                      //   mode

    // "Not a number" was input
    input         iSNaNS5);

    wire   [47:0] Man2Shifted;        // shifted
                                      //   mantissa

    wire   [48:0] ManS2Norm, ManS2SubnormalNorm; // normalized
                                                  //   mantissa

    // amount shifted during normalization
    wire   [7:0]  NormShiftedS2;
    // number of times to shift mantissa if result is sub-normal
    wire   [7:0]  ShiftSubNormalS2;
    // Value to be subtracted for rounding adjustment
    wire          RoundAdjS2;
    // amount to shift second mantissa
    wire   [7:0]  ShiftVal;
    // input mantissas in stage 0
    wire   [23:0] Man1, Man2;
    wire          IneAddS0;
    wire   [22:0] ManS4p1;
    // adjusted exponent in stage 3
    wire   [7:0]  ExpOutS4;
    wire          coS4;
```

```verilog
// staged sign bits
reg             Sign1S2, Sign2S2, SignS3, SignS4, SignS5;
// normalized mantissa
reg     [48:0] ManS3Norm, ManS3SubnormalNorm, ManSubNormAdjS3;
// staged exponent bits
reg     [7:0]  ExpS2, ExpS3, ExpS4;
// staged mantissa values
reg     [48:0] ManS2, ManS3, ManS4, ManS5;
// staged inexact flags
reg             IneAddS1, IneAddS2, IneAddS3, IneAddS4, IneAddS5;
// staged rounding mode values
reg     [1:0]  RModeS0, RModeS1, RModeS2, RModeS3, RModeS4,
    RModeS5;

// flags if operation was subtract
reg             SubS2, SubS3, SubS4;
// flags if mantissa is zero
reg             ManZeroS3;
// input sign
reg             s1, s2;
// input exponent
reg     [7:0]  e1, e2;
// input mantissa
reg     [22:0] f1, f2;
// flags a round-down adjustment
reg             ExpAdjS4;
reg             ManZeroS2;
reg     [7:0]  NormShiftedS3;
// shifted mantissa
reg     [47:0] Man2ShiftedS1;
reg     [23:0] Man1S1;
// adjusted floating point inputs: swap so first is larger
reg     [31:0] FP1Adj, FP2Adj;
// mantissa stage 3, and man stage 3 plus 1
reg     [22:0] ManS5Out, ManS5p1;
// carry out from plus 1 operation
reg             coS5;
// adjusted exponent in stage 3
reg     [7:0]  ExpOutS5;
// flags a swap of input operators
reg             SwapS0;
reg     [7:0]  ShiftValSwapS0, ShiftValNoSwapS0;
reg             RoundAdjS3;

assign  Man1            = (FP1Adj[30:23] == 0) ? {1'b0,
                          FP1Adj[22:0]} :
                            // if e1=0, then it is subnormal
                            {1'b1, FP1Adj[22:0]};
```

```
assign   Man2             = (FP2Adj[30:23] == 0) ? {1'b0,
                            FP2Adj[22:0]} :
                            {1'b1, FP2Adj[22:0]};

// shift less 1 if the smaller value is subnormal and
   larger is not
assign   ShiftVal         = SwapS0 ? ShiftValSwapS0 :
                            ShiftValNoSwapS0;

// stage 3 mantissa plus 1
assign {coS4, ManS4p1} = ManS4[47:25] + 1;

assign ExpOutS4           = ExpS4 - ExpAdjS4;

// adjust for rounding mode
// - if rounding away from infinity, then we end up at
   largest value
assign oFPSum             = ((ExpOutS5 == 8'hff) & !iSNaNS5 &
                            (RModeS5 == 'ROUNDEVEN )) ?
                            {SignS5, 'INF } :
                            ((ExpOutS5 == 8'hff) & !iSNaNS5 &
                            (RModeS5 == 'ROUNDZERO )) ?
                            {SignS5, 'LARGESTVAL } :
                            ((ExpOutS5 == 8'hff) & !iSNaNS5 &
                            (RModeS5 == 'ROUNDUP ) & !SignS5) ?
                            {1'b0,'INF } :
                            ((ExpOutS5 == 8'hff) & !iSNaNS5 &
                            (RModeS5 == 'ROUNDUP ) & SignS5) ?
                            {1'b1, 'LARGESTVAL } :
                            ((ExpOutS5 == 8'hff) & !iSNaNS5 &
                            (RModeS5 == 'ROUNDDOWN ) & !SignS5) ?
                            {1'b0,   'LARGESTVAL } :
                            ((ExpOutS5 == 8'hff) & !iSNaNS5 &
                            (RModeS5 == 'ROUNDDOWN ) & SignS5) ?
                            {1'b1,'INF } :
                            ((ExpOutS5 == 8'hff) && iSNaNS5) ?
                            {SignS5, ExpOutS5, 1'b1,
                            ManS5Out[21:0]} :
                            {SignS5, ExpOutS5 + (coS5 &
                            (ManS5Out == ManS5p1)), ManS5Out};
                            // adjust exponent if there is a
                              carry over

// overflow if we reached our maximum value
assign oOverflowAdd       = ((oFPSum[30:0] == 'LARGESTVAL) &
                            ({ExpOutS5, ManS5Out} !=
                            'LARGESTVAL) );

// inexact also if there is overflow or if there are
   truncated bits
assign oIneAdd            = IneAddS5 | oOverflowAdd |
                            (|ManS5[24:0]);
```

```
// number of times to shift mantissa if result is subnormal
assign ShiftSubNormalS2 = ExpS2 + (ExpS2 == 0); // shift at
                                                           least
                                                           once

// Rounding conditions to subtract 1 from result
assign RoundAdjS2        = ((RModeS2 == 'ROUNDZERO ) &
                           IneAddS2 & SubS2 & ManZeroS2) |
                           ((RModeS2 == 'ROUNDDOWN ) &
                           IneAddS2 &
                           !Sign1S2 & SubS2 & ManZeroS2) |
                           ((RModeS2 == 'ROUNDUP ) &
                           IneAddS2 & Sign1S2 & SubS2 &
                           ManZeroS2);

// pre-normalize second operator so that decimals are
//    aligned
PreNormAdd PreNormAdd    (.iPreShift(Man2),
                         .iShiftVal(ShiftVal),
                         .oPostShift(Man2Shifted),
                         .oIneAdd(IneAddS0));

// normalize result by shifting mantissa and adjusting
// exponent by NormShifted
PostNormAdd PostNormAdd (.iMantissa(ManS2),
                         .oMantissaNorm(ManS2Norm),
                         .oNormShifted(NormShiftedS2));

// normalization if result is sub normal
NormSubNormalAdd NSNA    (.iPreShift(ManS2),
                         .oPostShift(ManS2SubnormalNorm),
                         .iShiftVal(ShiftSubNormalS2));

always @(posedge iClk) begin
  // Stage 0
  // First FP must be bigger than the second
  // if not, then swap the two
  if(iFP1[30:0] > iFP2[30:0]) begin
    FP1Adj              <= iFP1;
    FP2Adj              <= iFP2;
    SwapS0              <= 0;
  end
  else begin
    FP1Adj              <= iFP2;
    FP2Adj              <= iFP1;
    SwapS0              <= 1;
  end
  ShiftValNoSwapS0      <= iFP1[30:23]-iFP2[30:23] -
                          ((iFP2[30:23] == 0) &
                          (iFP1[30:23] != 0));
```

```
ShiftValSwapS0        <= iFP2[30:23]-iFP1[30:23] -
                         ((iFP1[30:23] == 0) &
                         (iFP2[30:23] != 0));

RModeS0 <= iRMode;

// Stage 1
{s1, e1, f1}          <= FP1Adj; // pick out fields from
                                    raw FP values
{s2, e2, f2}          <= FP2Adj;
RModeS1               <= RModeS0;
IneAddS1              <= IneAddS0;
Man2ShiftedS1         <= Man2Shifted;
Man1S1                <= Man1;

// Stage 2
Sign1S2               <= s1;
Sign2S2               <= s2;
ExpS2                 <= e1;
RModeS2               <= RModeS1;
IneAddS2              <= IneAddS1;
ManZeroS2             <= (Man2ShiftedS1 == 0); // flags
                                                  addition
                                                  to zero

// add or sub mantissa values
if(s1 == s2) begin // add mans if signs equal
  ManS2               <= {Man1S1, 24'b0} + Man2ShiftedS1;
  SubS2               <= 0;
end
else begin   // subtract mans if signs opposite
  ManS2               <= {Man1S1, 24'b0} - Man2ShiftedS1;
  SubS2               <= 1;
end

// Stage 3
SignS3                <= Sign1S2;
ExpS3                 <= ExpS2;
IneAddS3              <= IneAddS2;
RModeS3               <= RModeS2;
ManZeroS3             <= ManZeroS2;
SubS3                 <= SubS2;
ManS3                 <= ManS2;
ManS3Norm             <= ManS2Norm;
ManS3SubnormalNorm    <= ManS2SubnormalNorm;
NormShiftedS3         <= NormShiftedS2;
ManSubNormAdjS3       <= ManS2SubnormalNorm - RoundAdjS2;
RoundAdjS3            <= RoundAdjS2;

// Stage 4
RModeS4               <= RModeS3;
```

```
// zeroth bit shifted out of mantissa - if 1 then inexact
IneAddS4                <= IneAddS3;
SubS4                   <= SubS3;

if(ManS3 == 0) begin
  // sign depends on rounding mode
  SignS4                <= ((RModeS3 == 'ROUNDDOWN) &
                          (SubS3 | SignS3)) |
                          ((RModeS3 == 'ROUNDEVEN) &
                          (!SubS3 & SignS3)) |
                          ((RModeS3 == 'ROUNDZERO) &
                          (!SubS3 & SignS3)) |
                          ((RModeS3 == 'ROUNDUP) &
                          (!SubS3 & SignS3));

  // if the total mantissa is zero, then result is zero
  // and therefore exponent is zero
  ExpS4                 <= 0;
  ExpAdjS4              <= 0;
  ManS4                 <= ManS3Norm;  // normalized result
end
else if((ExpS3 < NormShiftedS3) & (NormShiftedS3 != 1)) begin
  // the result is a subnormal number
  SignS4                <= SignS3;
  ExpS4                 <= 0;
  ExpAdjS4              <= 0;
  ManS4                 <= ManSubNormAdjS3; // adjust for
                                            rounding mode
end
else begin
  // otherwise, the final exponent is reduced by the
     number of
  // shifts to normalize the mantissa
  SignS4                <= SignS3;
  ExpS4                 <= ExpS3 - NormShiftedS3 + 1 +
                          (ExpS3 == 0);
  ExpAdjS4              <= (RoundAdjS3 &
                          (ManS3Norm[47:24] == 0));
  ManS4                 <= ManS3Norm - {24'b0, RoundAdjS3,
                          24'b0};
end

// Stage 5
SignS5                  <= SignS4;
RModeS5                 <= RModeS4;
ManS5p1                 <= ManS4p1;
IneAddS5                <= IneAddS4;
ManS5                   <= ManS4;
coS5                    <= coS4;
```

Table 9.2   Implementation in a Xilinx Spartan-3

Frequency	65 MHz	110 MHz
FFs	538	1087
LUTs	2370	2363

```
// adjust for rounding mode
// - various conditions to round up
ManS5Out              <= (RModeS4 == 'ROUNDEVEN &((ManS4[24]&
                      |ManS4[23:0])|(ManS4[24] & ManS4[25]&
                      ((SubS4 & !IneAddS4)|!SubS4)))?
                      ManS4p1:
                      (RModeS4 == 'ROUNDUP)&((|ManS4[24:0]&
                      !SignS4)|(IneAddS4 & !SignS4 &
                      !SubS4))?
                      ManS4p1 : (RModeS4 == 'ROUNDDOWN ) &
                      ((|ManS4[24:0] & SignS4) | (IneAddS4 &
                      SignS4 & !SubS4)) ? ManS4p1 :
                      ManS4[47:25];

ExpOutS5              <= ExpOutS4;
end
endmodule
```

9.2.2   Resources and Performance

This section reports the resource utilization of the Add-Sub module targeted at various architectures as well as performance measurements. The pipelined architecture is designed for maximum throughput and therefore has many pipelining registers and parallel logic. The implementation results are shown in Table 9.2.

The throughput at the maximum frequency is 3.52 Gbps. Note that we can avoid register duplication by targeting a slower frequency. At the low end, we can achieve a throughput of 2.08 Gbps.

Much of the area in a true IEEE floating-point calculation is related to the subnormal region, overflow conditions, and the detection of various conditions of the input/output including infinity, not-a-number, and so forth. If a given application does not have the requirement to be IEEE compliant and the range can be bounded within the normal region of the floating-point representation, the design can be implemented with about half the resources shown in Table 9.2.

Chapter 10

Reset Circuits

Despite their critical importance, reset circuits are among the most often ignored aspects of an FPGA design. A common misconception among FPGA designers is that reset synchronization is only important in ASIC design and that the global reset resources in the FPGA will handle any synchronization issues. This is simply not true. Most FPGA vendors provide library elements with both synchronous and asynchronous resets and can implement either topology. Reset circuits in FPGAs are critically important because an improperly designed reset can manifest itself as an unrepeatable logical error. As mentioned in previous chapters, the worst kind of error is one that is not repeatable.

This chapter discusses the problems associated with improperly designed resets and how to properly design a reset logic structure. To understand the impact reset has on area, see Chapter 2.

During the course of this chapter, we will discuss the following topics:

- Discussion of asynchronous versus synchronous resets.

  Problems with fully asynchronous resets
  Advantages and disadvantages of fully synchronous resets
  Advantages of asynchronous assertions, synchronous deassertion of reset
- Issues involved with mixing reset types.

  Flip-flops that are not resettable
  Dealing with internally generated resets
- Managing resets over multiple clock domains.

Advanced FPGA Design. By Steve Kilts
Copyright © 2007 John Wiley & Sons, Inc.

10.1 ASYNCHRONOUS VERSUS SYNCHRONOUS

10.1.1 Problems with Fully Asynchronous Resets

A fully asynchronous reset is one that both asserts and deasserts a flip-flop asynchronously. Here, *asynchronous reset* refers to the situation where the reset net is tied to the asynchronous reset pin of the flip-flop. Additionally, the reset assertion and deassertion is performed without any knowledge of the clock. An example circuit is shown in Figure 10.1.

The code for the asynchronous reset of Figure 10.1 is trivial:

```
module resetff(
  output reg  oData,
  input       iClk, iRst,
  input       iData);

  always @(posedge iClk or negedge iRst)
    if(!iRst)
      oData <= 0;
    else
      oData <= iData;
endmodule
```

The above coding for a flip-flop is very common but is very dangerous if the module boundary represents the FPGA boundary. Reset controllers are typically interested in the voltage level they are monitoring. During power-up, the reset controller will assert reset until the voltage has reached a certain threshold. At this threshold, the logic is assumed to have enough power to operate in a valid

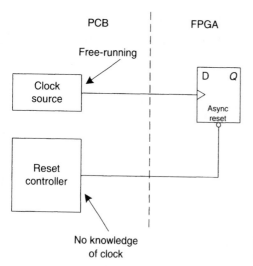

Figure 10.1 Example asynchronous reset source.

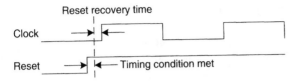

Figure 10.2 Reset recovery time.

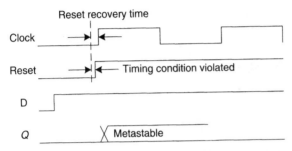

Figure 10.3 Reset recovery time violation.

manner, and so the reset is deasserted. Likewise during a power-down or a brown-out condition, reset is asserted when the voltage rail drops below a corresponding threshold. Again, this is done with no thought to the system clock of the device that is being reset.

The biggest problem with the circuit described above is that it will work most of the time. Periodically, however, the edge of the reset deassertion will be located too close to the next clock edge and violate the reset recovery time. The reset recovery time is a type of setup timing condition on a flip-flop that defines the minimum amount of time between the deassertion of reset and the next rising clock edge as shown in Figure 10.2.

As can be seen in the waveform of Figure 10.2, the reset recovery condition is met when the reset is deasserted with an appropriate margin before the rising edge of the clock. Figure 10.3 illustrates a violation of the reset recovery time that causes metastability at the output and subsequent unpredictable behavior.

Reset recovery time violations occur at the deassertion of reset.

It is important to note that reset recovery time violations only occur on the deassertion of reset and not the assertion. Therefore, fully asynchronous resets are not recommended. The solutions provided later in this chapter regarding the reset recovery compliance will be focused on the transition from a reset state to a functional state.

10.1.2 Fully Synchronized Resets

The most obvious solution to the problem introduced in the preceding section is to fully synchronize the reset signal as you would any asynchronous signal. This is illustrated in Figure 10.4.

The code to implement a fully synchronous reset is similar to the double-flopping technique for asynchronous data signals.

```
module resetsync(
    output reg   oRstSync,
    input        iClk, iRst);
    reg          R1;

    always @(posedge iClk) begin
    R1           <= iRst;
    oRstSync     <= R1;
    end
endmodule
```

The advantage to this type of topology is that the reset presented to all functional flip-flops is fully synchronous to the clock and will always meet the reset recovery time conditions assuming the proper buffering is provided for the high fan-out of the synchronized reset signal. The interesting thing about this reset topology is actually not the deassertion of reset for recovery time compliance as discussed in the previous section but rather the assertion (or more specifically the duration of reset). In the previous section, it was noted that the assertion of reset is not of interest, but that is true only for asynchronous resets and not necessarily with synchronous resets. Consider the scenario illustrated in Figure 10.5.

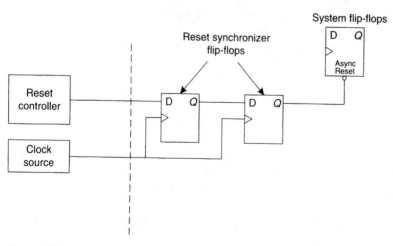

Figure 10.4 Fully synchronized reset.

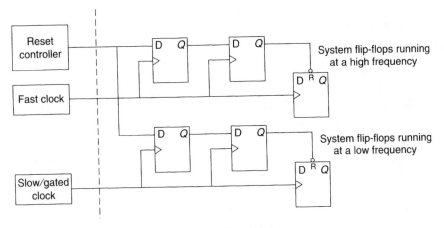

Figure 10.5 Fully synchronous reset with slow/gated clock.

Here, a single reset is synchronized to a fast clock domain and a relatively slow clock domain. For purposes of illustration, this could also be a non-periodic-enabled clock. What happens to the circuit's ability to capture the reset under conditions where the clock is not running? Consider the waveforms shown in Figure 10.6.

In the scenario where the clock is running sufficiently slow (or when the clock is gated off), the reset is not captured due to the absence of a rising clock edge during the assertion of the reset signal. The result is that the flip-flops within this domain are never reset.

Fully synchronous resets may fail to capture the reset signal itself (failure of assertion) depending on the nature of the clock.

For this reason, fully synchronous resets are not recommended unless the capture of the reset signal (reset assertion) can be guaranteed by design. Combining the last two reset types into a hybrid solution that asserts the reset asynchronously and deasserts the reset synchronously would provide the most reliable solution. This is discussed in the next section.

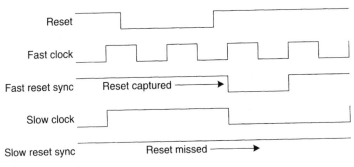

Figure 10.6 Reset synchronization failure.

10.1.3 Asynchronous Assertion, Synchronous Deassertion

A third approach that captures the best of both techniques is a method that asserts all resets asynchronously but deasserts them synchronously.

In Figure 10.7, the registers in the reset circuit are asynchronously reset via the external signal, and all functional registers are reset at the same time. This occurs asynchronous with the clock, which does not need to be running at the time of the reset. When the external reset deasserts, the clock local to that domain must toggle twice before the functional registers are taken out of reset. Note that the functional registers are taken out of reset only when the clock begins to toggle and is done so synchronously.

A reset circuit that asserts asynchronously and deasserts synchronously generally provides a more reliable reset than fully synchronous or fully asynchronous resets.

The code for this synchronizer is shown below.

```
module resetsync(
    output reg  oRstSync,
    input       iClk, iRst);
    reg         R1;

always @(posedge iClk or negedge iRst)
    if(!iRst) begin
        R1        <= 0;
        oRstSync  <= 0;
    end
```

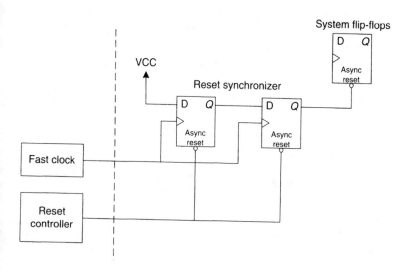

Figure 10.7 Asynchronous assertion, synchronous deassertion.

```
     else begin
        R1          <= 1;
        oRstSync    <= R1;
     end
  endmodule
```

The above reset implementation allows a group of flip-flops to be placed into reset independent of a clock but taken out of reset in a manner synchronous with the clock. As a matter of good design practice, the asynchronous assertion–synchronous deassertion method is recommended for system resets.

10.2 MIXING RESET TYPES

10.2.1 Nonresetable Flip-Flops

As a matter of good design practice, flip-flops of different reset types should not be combined into a single always block. The following code illustrates the scenario where a nonresetable flip-flop is fed by a resetable flip-flop:

```
module resetckt (
   output reg oDat,
   input       iReset, iClk,
   input       iDat);
   reg datareg;

always @(posedge iClk)
   if(!iReset)
      datareg   <= 0;
   else begin
      datareg   <= iDat;
      oDat      <= datareg;
   end
endmodule
```

The second flip-flop (oDat) will be synthesized with a flip-flop that has a load or clock enable input driven by the reset of the first. This is illustrated in Figure 10.8.

This requires larger sequential elements as well as extra routing resources. If the two flip-flops are split according to the following code:

```
module resetckt(
   output reg oDat,
   input       iReset, iClk);
   input       iDat);
   reg         datareg;

   always @(posedge iClk)
      if(!iReset)
         datareg <= 0;
```

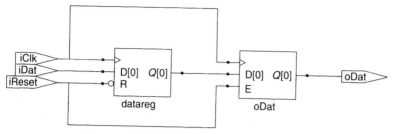

Figure 10.8 Implementation with mixed reset types.

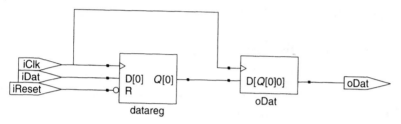

Figure 10.9 Optimal implementation with mixed reset types.

```
        else
            datareg <= iDat;
        always @(posedge iClk)
            oDat      <= datareg;
        endmodule
```

the second flip-flop will have no unnecessary circuitry as represented in Figure 10.9.

Different reset types should not be used in a single always block.

10.2.2 Internally Generated Resets

In some designs, there are conditions where an internal event will cause a reset condition for a portion of the chip. There are two options in this case:

- Using asynchronous resets, design the logic that derives the reset to be free of any static hazards.
- Use synchronous resets.

The main problem with a static hazard on an asynchronous reset is that due to variances in propagation delays, a reset pulse could occur even though the logic is switching from one inactive state to another. Assuming an active low

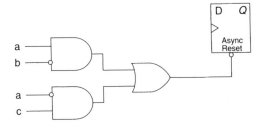

Figure 10.10 Potential hazard on reset pin.

reset, a glitch on an inactive state would be defined as a static-1 hazard. As an example, consider Figure 10.10.

With the circuit in Figure 10.10, a static-1 hazard can occur during the transition (a, b, c) = (1, 0, 1) −> (0, 0,1) as shown in Figure 10.11.

As can be seen from the waveforms in Figure 10.11, a static-1 glitch can occur on the reset line when one of the terms that sets the output inactive (logic-1) becomes invalid before the next term sets the reset high. Remembering back to logic design 101, this can be represented in a K-map format as shown in Figure 10.12.

Each circled region in Figure 10.12 indicates a product term that sets the reset inactive. The static-1 hazard occurs when the state of the inputs changes from one adjacent product term to another. If the first product term becomes inactive before the second is set, a glitch will occur. To fix this problem, a redundant prime implicant is created to bridge the two product terms as shown in Figure 10.13.

By adding the new product term as indicated by the redundant logic mapping, we eliminate the possibility of a hazard that will cause a glitch while the reset is in the inactive state.

In general, the technique for eliminating static-1 hazards will prevent a glitch on an internally generated reset line. However, the use of a fully synchronous

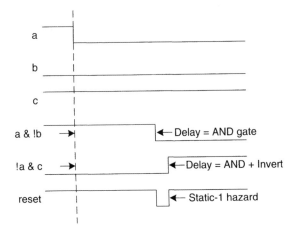

Figure 10.11 Example waveform with reset hazard.

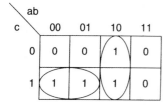

Figure 10.12   Identifying the static-1 hazard.

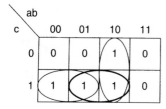

Figure 10.13   Adding a prime implicant.

reset synchronizer is typically recommended in practice. This helps to maintain a fully synchronous design and to eliminate the redundant logic necessary to maintain a glitch-free reset signal.

10.3 MULTIPLE CLOCK DOMAINS

We have already established that reset deassertion must always be synchronous and that asynchronous reset signals must be resynchronized for deassertion. As an

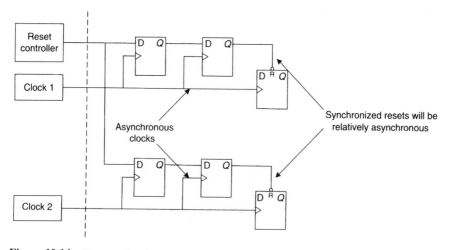

Figure 10.14   Reset synchronization with multiple clock domains.

extension of this principle, it is important to synchronize resets independently for each asynchronous clock domain.

As can be seen in Figure 10.14, a separate reset synchronization circuit is used for each clock domain. This is necessary because of the fact that a synchronous reset deassertion on one clock domain will not solve the clock recovery problems on an asynchronous clock domain. In other words, a synchronized reset signal will still be asynchronous relative to the synchronized reset from an independent clock domain.

A separate reset synchronizer must be used for each independent clock domain.

10.4   SUMMARY OF KEY POINTS

- Reset recovery time violations occur at the deassertion of reset.

- Fully synchronous resets may fail to capture the reset signal itself (failure of assertion) depending on the nature of the clock.

- A reset circuit that asserts asynchronously and deasserts synchronously generally provides a more reliable reset than fully synchronous or fully asynchronous resets.

- Different reset types should not be used in a single always block.

- A separate reset synchronizer must be used for each independent clock domain.

Chapter 11

Advanced Simulation

Because of recent advancements in rapid implementation, FPGA programming, and in-system debugging, many FPGA designers are spending less time creating comprehensive simulation testbenches and relying more on hardware debug to validate their designs. A trend with many modern FPGA designers is to only write "quick-and-dirty" simulations for individual modules, ignore the top-level sims, and anxiously jump directly to hardware. This is of course not the case in regulated industries such as medical or aviation but is the case in many of the thousands of new unregulated industries recently introduced to the power of FPGAs. Although in-system debug methods have become very sophisticated and design methods that focus on this type of debug and design validation have matured, one rarely regrets creating a comprehensive, fully automated simulation environment.

This chapter discusses many of the techniques involved in creating a useful simulation environment for verifying FPGA designs and describes a number of heuristics that have proved to work across numerous industries. During the course of this chapter, we will discuss the following topics:

- Architecting a testbench.
  Components of a testbench
  Proper flow of a testbench including the main thread, clock generation, and testcase inclusion
- Creating system stimulus using tools such as MATLAB.
- Bus-functional models for common interfaces.
- Gaining insight into overall coverage of stimulus.
- Running gate-level sims for verification, debug, and power estimation.
- Common testbench traps and proper methods for modeling devices.

Advanced FPGA Design. By Steve Kilts
Copyright © 2007 John Wiley & Sons, Inc.

11.1 TESTBENCH ARCHITECTURE

The first step in creating a useful simulation environment is setting up and organizing the testbench properly. The testbench is the top-level module in the simulation that is responsible for stitching all the models together. Often, the testbench will provide portions of the stimulus depending on the size and complexity of the design. A poorly designed testbench, typically one that was originally designed to be quick and dirty, can grow into a monstrous abomination of scattered behavioral constructs and stimulus that no one can read or completely understand.

11.1.1 Testbench Components

A top-level testbench can be modeled abstractly according to Figure 11.1.

The testbench is the top-level module in the simulation and stitches together all subcomponents in the system model. The test procedure, which typically resides within the testbench, manages the main thread of the simulation and the flow of tests. This process defines which tests are run, which vectors are used, and how the data is logged and reported.

The global stimulus represents the basic vectors that apply to the system as a whole. These most often include the system clocks, resets, or any initial conditions that set the system into an appropriate state for simulation.

The hardware models are called out in the testbench as well. These are the devices-under-test in the simulation, and these are the modules that will ultimately

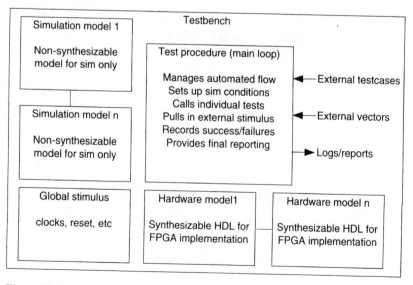

Figure 11.1 Testbench components.

be implemented in the FPGA. Often, there is only one hardware model in the testbench, that is, the top-level module in the FPGA.

Finally, the testbench calls out any number of simulation models. These modules represent other components in the system that the FPGA will interact with but will not be implemented in the FPGA. These could be models of memories, microprocessors, analog components, or any other system components that interact with the FPGA.

11.1.2 Testbench Flow

The flow of a simulation is typically quite a bit different from the hardware description itself. At a very fundamental level, the hardware is described in a concurrent fashion, whereas simulations operate in a procedural fashion. In other words, hardware is described based on logical operations triggered from synchronous events. There are no valid descriptions for chronological execution in synthesizable hardware and synthesis will simply ignore timing information written into the code. This is the primary conceptual difference between software and hardware design. Simulation code, similar to a software design, operates on primarily a procedural basis. In other words, a simulation will apply stimulus a, then b, then c in a particular order and with specific timing. Thus, simulation much more closely represents software design than hardware design.

11.1.2.1 Main Thread

To model procedural behavior in the testbench, certain behavioral constructs are used that would not be used for hardware design. In Verilog, the main simulation thread is often modeled with an initial statement that contains a series of blocking expressions. Consider the following code example.

```
initial begin
errors = 0; // reset error count

// reset inputs to chip
chipin1 = 0;
chipin2 = 16'ha5;

// reset simulation parameters
resetsim();

// reset for chip
reset_fpga();

//
// Add testcases here
//
'include "test1.v"
'include "test2.v"
```

```
$display("\nSimulation completed with %d errors\n", errors);
$stop;
end
```

There are a number of interesting observations that can be made from the above code. First, all assignments are blocking in nature. This means that they will be executed one at a time in a sequence, similar to execution in software. Second, the main simulation thread only contains global initialization assignments, whereas all specific functional tests are written into the individual testcases "test1.v" and "test2.v." Third, the main execution path is kept thin and modularized. In other words, the main initial statement does not contain large amounts of initialization data to maintain a readable format.

It is good design practice to partition the individual test cases from the main thread.

Finally, the testbench is self-checking. It is assumed that each of the individual functional tests comprehensively checks the conditions within the hardware models, reports any mismatches to the standard output or the log file, and increments the error count. When the simulation completes, a success or failure is indicated by the final statement. This basic automation greatly reduces the amount of time spent verifying vectors.

It is good design practice to create an automated, self-checking testbench. This will save a significant amount of time down the road as the testbench grows.

There is a certain temptation to skip this automation step early on in the development of the testbench. However, this typically creates major portability issues downstream as new designers begin to analyze the system. If a new designer is focusing on a particular module, it may be necessary to verify particular changes against the system to ensure that nothing else was broken by the modifications. A fully automated testbench will greatly speed this process as it will allow for the appropriate validation without complete understanding of the system.

11.1.2.2   Clocks and Resets

The testbench clocks and resets should be modeled at a global level. The main initial statement should not define the starting or transition points for the clocks and resets. Instead, an independent initial statement external to the main loop generates the initial condition and the oscillation.

```
`timescale 1ns/1ns

`define PERIOD 5  // 100MHz clock

initial begin
clk <= 0;
forever #(`PERIOD) clk = ~clk;
end
```

In the above code, the timescale is set to 1 ns meaning that any timing information as defined by the "#" symbol will be of the order 1 ns. Also, by defining the period as 5 ns, we are really defining the time between transitions on the clock (there will be two transitions in a 10-ns period). In the above example, the signal clk will be logic-0 at time = 0, logic-1 at time = 5 ns, logic-0 at time = 10 ns, and so on throughout the entire simulation.

The reset can be modeled in a similar fashion as shown below.

```
initial begin
  reset <= 0;
  @(posedge clk); //may need several cycles for reset
  @(negedge clk) reset = 1;
end
```

One interesting thing to note about these initial blocks is that we have mixed non-blocking and blocking assignments, noted in later chapters to be a bad coding style. Although it is certainly a poor coding style for synthesizable always blocks, it is not necessarily so for initial blocks that are used for simulation only. In this case, nonblocking assignments to the initial value of logic-0 are used. This ensures that all always blocks in the design are evaluated prior to this assignment. In other words, this ensures that any always blocks that might be triggered from the initial assignment are properly evaluated. If asynchronous resets are used, then this would apply to the initial reset assertion as well (assuming it is asserted at time 0).

> Initialize testbench clocks and resets with nonblocking assignments and update them with blocking assignments.

11.1.2.3 Test Cases

If at all possible, it is good design practice to make the test cases themselves as modular as possible. By this we mean that any of the test-case calls can be removed from the testbench, reordered, or new test cases inserted. This is important as it clearly defines a boundary for the conditions for a particular test making the simulation more understandable. Also, during the debug phase it will be very common to run only particular tests to attempt to reproduce a problem.

> Create test cases such that they can stand alone inside the main thread.

If the entire test suite is extensive and time consuming, it will be desirable to quickly remove all unnecessary tests that do not focus on the problem at hand. This is where a modular design becomes extremely useful.

Consider the following code for the first test case defined in our testbench:

```
// test1.v
// test scenario #1
resetsim();
reset_fpga();
```

```
$display("Begin testing scenario 1... \n");
...
// compare output with expected value and report if mismatch
verify_output(output_value, expected_value);

$display("\nCompleted testing scenario 1 with %d errors",
    errors);
```

There are a number of notable aspects to the example test case above. First, the test case ensures modularity by resetting the simulation parameters as well as resetting the FPGA. If every test case resets the simulation, modularity can be ensured. In most cases, this will have a performance impact on the simulation run-time, but as a matter of good design practice this is typically recommended. Second, note that the reset is always deasserted on the falling edge of the clock. Due to the fact that the free running clock may not be reset with the other simulation parameters, it is important to have a consistent method of releasing the reset to avoid any reset recovery times (although a good reset circuit will eliminate this problem). Third, a common function called "verify_output()" is used throughout all test cases as a means to compare simulated values with expected values. This function is typically defined in the testbench itself but used in the particular test cases. This checking function can be implemented as a task in Verilog:

```
task verify_output;
    input [23:0] simulated_value;
    input [23:0] expected_value;
    begin
    if (simulated_value[23:0] != expected_value[23:0])
        begin
        errors = errors + 1;
        $display("Simulated Value = %h, Expected Value = %h,
        errors = %d,
        at time = %d\n", simulated_value, expected_value,
        errors, $time);
        end
    end
endtask
```

The task shown above simply compares the simulated value with the expected value (both of which were passed in from the test case), and any mismatches are reported.

An additional note for creating test cases is to try and reference signals at module boundaries. In other words, when comparing signals inside the FPGA hierarchy to expected vectors, it is good design practice to reference signals that are defined as inputs or outputs somewhere in the hierarchy. The reason for this is that during the debug phase, a designer will often back-annotate a netlist into the simulation environment to debug a specific problem or to simply run the entire test suite against the final implementation (design that has been placed and routed) depending on the verification standards at a particular organization. When back-annotating a netlist with all low-level component and timing information, it is fairly easy to maintain the hierarchy but quite difficult to preserve the structure and naming conventions of all nets and registers. There is a particular conflict to preserving names if certain optimizations such as resource sharing or register balancing are performed by the synthesis tool.

Reference signals inside the design at the module boundaries whenever possible.

If the design is partitioned well and the hierarchy is preserved during implementation, then a testbench that references only module boundaries will have a much easier mapping to a back-annotated environment.

11.2 SYSTEM STIMULUS

Generating stimulus for a testbench can be one of the most tedious and time-consuming tasks of the simulation development. At a very low level (typically at the module level sim), it is easy to "bit-bang" the inputs with hard-coded values that exercise the basic functionality of a design. For instance, in the case of an FIR filter, it may be fairly straightforward to present a dozen or so inputs that test the basic multiply and accumulate functionality at the boundary conditions. However, at a system level, it is not so easy to generate stimulus and analyze the results to measure characteristics of the filter such as frequency and phase response. Also, when communicating with standardized interfaces such as PCI or an SDRAM device, it is no easy task to generate a set of hard-coded vectors that emulate the response of the device.

11.2.1 MATLAB

MATLAB is a high-level mathematical modeling and analysis tool that is extremely useful when generating large sets of vectors that have some regular pattern or mathematical description. For the case of the DSP filter described earlier, it would be difficult to generate the appropriate superposition of signals to allow for a proper filter simulation. With a tool like MATLAB, however, this is trivial.

```
clear;
% generate 100Hz wave sampled at 40kHz
sinewave = sin(10.*(1:4000).*2.*pi./4000); % 10 periods
sinehalfamp = sinewave * hex2dec('3fff'); % for normalization
fid = fopen('100hz.vec', 'w'); % file containing sim vectors
% normalize for 2's complement
for i=1:length(sinehalfamp)
    if(sinehalfamp(i) < 0)
      sinehex(i) = floor(hex2dec('ffff') + sinehalfamp(i) + 1);
    else
      sinehex(i) = floor(sinehalfamp(i));
    end
```

```
fwrite(fid, dec2hex(sinehex(i), 4));
fprintf(fid, '\n');
end
fclose(fid);
```

The MATLAB script above generates 10 periods of a 100-Hz sine wave sampled at 40 kHz (each vector represents a sample). The data is normalized to a 16-bit 2's complement hex format that can be read into a Verilog simulation with minimal effort.

MATLAB can be very useful when creating large or complex patterns for simulation.

11.2.2 Bus-Functional Models

Very often, certain system-level components are important to simulate yet difficult to model. An HDL netlist or Spice model will provide a high level of accuracy and reliability, but it will suffer poor simulation performance. For off-the-shelf components such as memories, microprocessors, or even standard bus interfaces such as PCI, bus-functional models (BFMs) can provide a means to simulate against the interface defined in the model without requiring a model for the entire device. For instance, a bus-functional model for a PCI interface will provide all of the timing information and functional checks to verify that the PCI protocol is operating correctly. It will not, however, necessarily contain a model for a microprocessor or other device behind the PCI bus.

BFMs have an obvious advantage in their simplicity. A designer can verify against a set of timing or protocol requirements without simulating a low-level model for an entire device. The disadvantage is, of course, that the model is only as good as the engineer that writes it. It is typically recommended that a BFM is used when it is provided by the vendor of the device. For instance, consider the example where an FPGA interfaces the PCI bus with a burstable Flash device as shown in Figure 11.2.

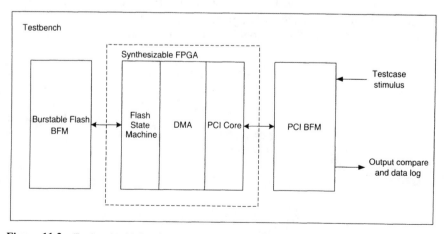

Figure 11.2 Testbench with bus functional models.

In this example, the synthesizable FPGA design sits between the BFMs for the PCI bus interface and the burstable Flash. The BFMs are simply high-level descriptions of the interfaces that will check the timing and protocol of the PCI core and the state machine to control the Flash. Most PCI core vendors such as Quicklogic will provide BFMs for the PCI bus, which allows the user to apply read and write commands at the testbench level. Additionally, vendors that sell large Flash devices with any complexity, such as Intel, will provide BFMs to verify the handshaking, buffering, and so forth.

11.3 CODE COVERAGE

A very powerful simulation technique is that of code coverage. To put it simply, code coverage is a feature of the simulation tool that provides statistical information regarding the simulation including the structures and transitions that were exercised and how often they were exercised. The most common pieces of data include lines that were covered (covered vs. not covered) with an associated "code coverage" number, state-machine coverage to verify all states and transitions, toggle coverage to estimate the amount of activity in different regions of the design, and so on. Code coverage is often used in an ASIC design because of the high importance verification plays in the process (every spin of the ASIC is relatively expensive, and thus the ASIC must be exhaustively simulated prior to tape-out).

Due to the strides forward in recent years relative to the sizes and densities of high-end FPGAs, designs have become complex enough that code coverage is playing a greater role in the verification of these as well. It is often not good enough to run a quick-and-dirty simulation and then debug in hardware. The designs are becoming too complex and the debug methods too time consuming. Code coverage provides a slick means to quickly determine which portions of the design have not been simulated. This can indicate weaknesses in the design and help to identify new verification tasks.

Code coverage checks the extent to which the design has been simulated and identifies any unsimulated structures.

11.4 GATE-LEVEL SIMULATIONS

The utility of gate-level simulations in the typical design flow has always been a debate, particularly in recent years with the advancements of in-system debugging tools such as Chipscope from Xilinx or Identify from Synplicity. Despite the speed at which these tools provide visibility into a design real-time, they cannot provide all the hooks necessary to observe every element and every net with characterized accuracy. Additionally, gate level simulations are the only way to debug certain types of unusual behavior, especially during conditions such as reset deassertion that are not easily analyzed by static timing analysis or in-system debug tools.

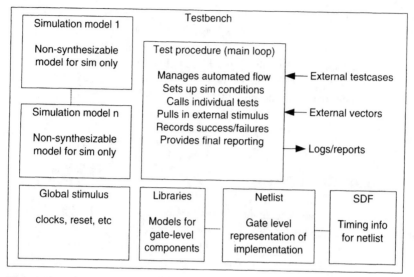

Figure 11.3   Testbench with gate-level blocks.

A well-designed testbench will typically allow for an easy migration to a gate-level simulation. Figure 11.3 illustrates the various components in a gate-level simulation.

Note that the simulation models, test procedures, global elements, external vectors, and so on are essentially unchanged from the RTL simulation environment. The primary change is the removal of the synthesizable RTL files and the addition of a netlist with supporting libraries and timing info.

The netlist is the gate-level representation of the design after implementation. The netlist contains a list of primitive elements in the FPGA with relevant interconnect information. Relative to the logic and cycle-by-cycle functionality, it should be identical to the RTL description. For instance, consider the following Verilog module.

```
module andor (
  output oDat,
  input  iDat1, iDat2, iDat3);
  wire   ANDNET;

  assign ANDNET = iDat1 & iDat2;
  assign oDat = ANDNET | iDat3;
endmodule
```

This code simply combines the three inputs into an AND-OR operation. Although we think of this as a two-gate operation, the low-level FPGA structure allows us to combine this entire operation into a single LUT. The physical implementation into a Xilinx device is represented in the netlist as shown below.

```
X_LUT4 #(.INIT(16'hFCF0)) oDat1 (
  .ADR0(VCC),
  .ADR1(iDat2_IBUF_1),
  .ADR2(iDat3_IBUF_2),
  .ADR3(iDat1_IBUF_0),
  .O(oDat_OBUF_3)
);
```

As can be seen from the LUT instantiation above, all I/O are connected to a single LUT that handles the two-logic-gate operation. Larger designs will have many LUT instantiations with wire interconnects to represent the physical implementation of the logic. Note that the LUT is configured (via the INIT statement) such that the logical operation is identical to our RTL description.

The library files are low-level technology models provided by the FPGA vendor that define the cells in the FPGA. In our Xilinx implementation above, the library file would contain a definition of the X_LUT4 element. Note that we cannot simulate this netlist until all cells are defined down to a level of abstraction that only contains generic Verilog constructs. The library file for the Xilinx cell shown above contains the following definition.

```
o_out = INIT[{a3, a2, a1, a0}];
```

Here, the signals a0–a3 are the inputs to the LUT, and INIT is a 16-element array of bits defined by the defparam statement in the netlist. Because Xilinx is an SRAM-based FPGA, all logic is packed into small LUTs implemented as SRAMs. The above description of the LUT is consistent with what we understand about how this works and is constructed with generic Verilog constructs that can be interpreted by any simulator.

The final component to the back-annotated netlist is the timing information. In the ASIC design world, a designer would be concerned about both hold and setup violations in the fast and slow extremes, respectively. However, in the FPGA world we are typically not concerned with hold delays due to the built-in signal delays through the routing matrices and the low-skew clock lines. Thus, we will only concern ourselves with the slow corner; that is, the extreme condition where the voltage supply is at the minimum level and temperature is at the maximum. The following entries in the SDF (Standard Delay Format) file illustrate this condition.

```
(VOLTAGE 1.14)
(TEMPERATURE 85)
(TIMESCALE 1 ps)
```

The above entries indicate that the timing defined in this SDF file are with the assumption that we want to simulate at the minimum voltage allowable (1.14 V) and the maximum temperature that this device is rated for (85°C). Additionally, all timing information is defined in picoseconds. The timing info for the single LUT is shown as follows.

```
(CELL (CELLTYPE "X_LUT4")
  (INSTANCE oDat1)
    (DELAY
      (ABSOLUTE
        (PORT ADR1 ( 499 ))
        (PORT ADR2 ( 291 ))
        (PORT ADR3 ( 358 ))
        (IOPATH ADR0 O ( 519 ))
        (IOPATH ADR1 O ( 519 ))
        (IOPATH ADR2 O ( 519 ))
        (IOPATH ADR3 O ( 519 ))
      )
    )
)
```

This SDF entry contains both interconnect and device delay. The SDF specification allows interconnect delay to be abstracted as a port delay; that is, only occurring at the input port of a device. The PORT entries of this LUT contain the interconnect delay from the input buffers to the individual inputs of this LUT. The IOPATH entries contain the timing through the LUT. Specifically, this is the delay from an input to the output and is defined here to be 519 ps.

A simulation with all of these components will provide a very accurate model of the inner workings of the FPGA as real vectors are used to stimulate the design. Not only can this be useful when debugging the design, but it can also be used to uncover other important pieces of information about the design as shown in the following sections.

11.5 TOGGLE COVERAGE

Node coverage is not frequently applied to gate-level netlists for FPGA design verification. Due to the ability of an FPGA to rapidly prototype the design and to run vectors real-time in hardware, the development time for gate-level code coverage is rarely justified. That said, code coverage can still be useful at the gate level as it is at the RTL level. When running a gate-level simulation, the code coverage analysis will provide a coverage number relative not to the logical constructs but to the physical elements themselves.

One aspect of code coverage at the gate level that can often provide useful information not available through a rapid prototype is that of toggle coverage. Toggle coverage provides not only coverage statistics in the form of "covered" and "not-covered" but also in the form of frequency-related information. In the absence of a more sophisticated power analysis tool, designers can use toggle coverage to produce an estimate of power dissipation based on the average number of toggles per clock cycle, the system voltage, and the average gate capacitance.

$$P_{dynamic} = CV_{dd}^2 f \qquad (11.1)$$

In the dynamic power-dissipation equation (11.1), statistical values for C (capacitance) and V_{dd} (supply voltage) are provided by the silicon manufacturer, and the frequency f is dependent on both the design and the stimulus. An accurate value for f is difficult to find without specific statistical information from the simulation environment.

To see how different stimulus will affect the dynamic power, the Xilinx tools can generate a testbench shell for the design that can be used as a starting point.

```
`include "C:/Xilinx/verilog/src/glbl.v"
`timescale 1 ns/1 ps

// Xilinx generated testbench shell
module testtb;
  reg iDat1;
  reg iDat2;
  reg iDat3;
  wire oDat;
  test UUT (
    .iDat1 (iDat1),
    .iDat2 (iDat2),
    .iDat3 (iDat3),
    .oDat  (oDat)
  );
  initial begin
    $display("    T iiio");
    $display("    i DDDD");
    $display("    m aaaa");
    $display("    e tttt");
    $display("      123 ");
    $monitor("%t",$realtime,,iDat1,iDat2,iDat3,oDat);
  end
  initial begin
    #1000 $stop;
    // #1000 $finish;
  end
endmodule
```

The testbench shell above includes the library file for global elements, instantiates the design under consideration (the instance of "test" is automatically assigned the name "UUT"), and initializes a text-based log of the vector changes. We can initialize the data and provide stimulus for periodic events.

```
initial begin
$dumpfile("test.vcd");
$dumpvars(1, testtb.UUT);
$dumpon;

#1000 $dumpoff;
#10 $stop;
end

initial begin
  iDat1 <= 0;
  forever #5 iDat1 = ~iDat1;
end

initial begin
  iDat2 <= 0;
  forever #10 iDat2 = ~iDat2;
end

initial begin
  iDat3 <= 0;
  forever #15 iDat3 = ~iDat3;
end
```

In the above simulation, all inputs are initialized to logic-0 and at 10-ns iDat3 transitions to a logic-1, which means the output will subsequently change to a logic-1. The "dump" commands define how the simulation vectors will be recorded in the VCD (vector change dump) file. This VCD file (test.vcd) contains the stimulus in vector format and can be used to estimate toggle activity and power estimation.

Gate-level simulations can be useful when estimating dynamic power dissipation.

Side Note: In the FPGA design world, more sophisticated power-estimation tools are often bundled with the implementation tools. Xilinx provides a tool called XPower, which provides power estimation for a given netlist and stimulus. After reading in the above VCD file, XPower provides the following as a dynamic power estimate.

```
Power summary:                                I(mA)      P(mW)
---------------------------------------------------------------
Total estimated power consumption:                        36
                                 ------
                        Vccint 1.20V:           6          7
                        Vccaux 2.50V:           7         18
                        Vcco25 2.50V:           5         12
```

The numbers provided by XPower will be similar to those generated by running toggle coverage and plugging the average toggle rate into our equation for dynamic power dissipation.

11.6   RUN-TIME TRAPS

11.6.1   Timescale

The timescale directive defines the units of the simulation timing as well as the resolution of the simulation. Consider the following timescale directive:

```
'timescale 1ns/100ps
```

This directive defines the absolute timing unit to be 1 ns, and the precision is defined to be 100 ps. The following assignment will resolve as 1.1 ns:

```
assign #1.1 out = in;
```

If the units were set to 100 ps and the resolution set to 10 ps, the delay in the above statement would resolve to 110 ps. One potential hazard in simulation is not providing sufficient resolution to the simulator. If an RTL simulation is performed with no back-annotated timing information and the system clock has a period that is an integer multiple of the timescale, the precision will have essentially no effect on the results. However, if fractional timing is defined (as it often is with SDF timing information), the resolution must be set high enough to resolve the number. If the resolution were set to 1 ns in the first example, the 1.1 ns delay would resolve to 1 ns and may cause simulation errors.

In addition to hazards due to resolutions that are too coarse, resolutions that are too fine will directly impact the simulation time. If the resolution is chosen to be, say, 1 ps when the finest timing precision is 100 ps, the simulation will run much slower than necessary. Because simulation run-time is often long (particularly with back-annotated netlists), this dramatic decrease in speed will directly affect the designer's productivity.

> Timescale precision must be chosen to balance simulation accuracy against run-time.

It is also important to note that although the timescale is clearly an important directive for simulations, it is completely ignored by synthesis tools. Thus, it is important that any absolute timing information only be used in the testbench and simulation-only modules and should not define specific behavior of synthesizable hardware.

11.6.2   Glitch Rejection

A common trap with gate-level simulations is the automatic glitch rejection built into simulation algorithms. A device is defined as having *transport delay* if a pulse of any width is propagated from the input to the output (Fig. 11.4).

A wire is a device that exhibits transport delay. There is no minimum pulse width required to pass over a wire model. In contrast, a device is defined as

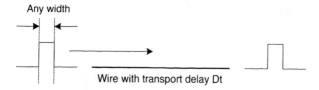

Figure 11.4   Transport delay.

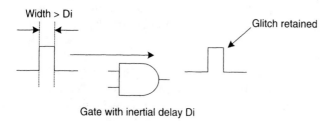

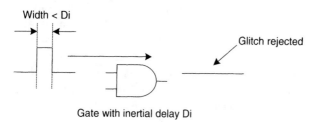

Figure 11.5   Inertial delay.

having *inertial delay* if there is a minimum pulse width required to propagate from a device input to the output.

A logic gate is a device that exhibits inertial delay. As can be seen from Figure 11.5, if the pulse width of the input to the logic gate is smaller than the inertial delay of the gate, the pulse (glitch) will be rejected at the output. This illustrates the importance of obtaining correct simulation models and providing accurate stimulus to the gate-level simulation. This is especially important if the logic is driving an asynchronous circuit that does not have the glitch-filtering benefit of an intermediate register.

Glitch rejection becomes an issue particularly when a designer needs to model delay through logic elements. The next section discusses the topic of proper delay modeling.

11.6.3   Combinatorial Delay Modeling

Delay modeling is a common problem for creating behavioral models of nonsynthesizable elements and interfaces. Typically, delays should not be added to synthesizable constructs to fix race conditions or to change the cycle-by-cycle

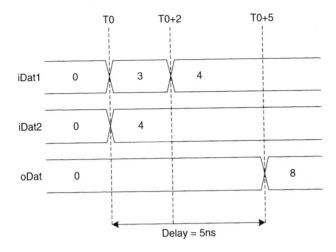

Figure 11.6   Result of incorrect delay modeling.

behavior of the circuit. Delays are always ignored by synthesis tools, and this type of modeling can easily create mismatches between simulation and synthesis. That said, hard-coded delays can be useful for simulation-only models to approximate certain types of behavior.

Due to the nature of the Verilog language, combinatorial delays should never be added to blocking assignments. Consider the following code for an 8-bit adder:

```
// INCORRECT DELAY MODELING
module delayadd(
  output reg [8:0] oDat,
  input [7:0] iDat1, iDat2);

  always @*
    #5 oDat = iDat1 + iDat2;
endmodule
```

The danger here is that a very simple simulation will probably work. If both inputs change at T0, the output will change at T0 + 5. However, with this type of modeling scheme, the output will change 5 ns after the first trigger of the always block since the last update to the output. In other words, if iDat1 and iDat2 change at T0, iDat1 changes again at T0 + 2, and the output will update at T0+5 with the latest values of both iDat1 and iDat2. This is illustrated in Figure 11.6.

This does not model combinatorial logic with 5 ns of prop delay and was not likely the intended behavior. To model pure transport delays, a designer could use nonblocking assignments as follows.

```
// POOR CODING STYLE
// MODELS TRANSPORT DELAY
module delayadd(
```

```
output reg [8:0] oDat,
input [7:0] iDat1, iDat2);

// nonblocking assignment typically not used for
   combinatorial logic
always @*
   oDat <= #5 iDat1 + iDat2;
endmodule
```

As can be seen in Figure 11.7, this accurately models the transport delay scenario of the previous example.

When modeling transport delay behavior such as with behavioral or bus-functional models, this method is sufficient. However, when modeling the behavior

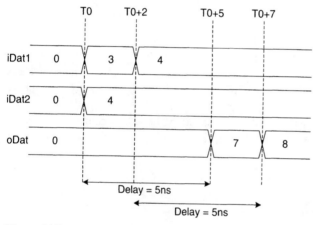

Figure 11.7 Result with transport delays.

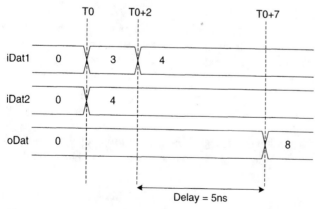

Figure 11.8 Result with inertial delays.

of true combinatorial logic, it is better to use a coding style that represents inertial delay. This is accomplished with a continuous assignment as shown below.

```
module delayadd(
  output [8:0] oDat,
  input [7:0] iDat1, iDat2);
  assign #5 oDat = iDat1 + iDat2;
endmodule
```

This type of representation will behave as shown in Figure 11.8.

As can be seen from the waveform of Figure 11.8, the output rejects the 2-ns pulse where iDat1 holds the value of 3. The output changes to the final value of 8 5-ns after both inputs have settled at 4. Thus, the continuous assignment models inertial delay and should be used for modeling combinatorial logic. If the combinatorial logic is complex and requires an always block, the outputs should drive a continuous assignment to model the delay.

Intertial delays due to combinatorial logic should be modeled with continuous assignments.

11.7 SUMMARY OF KEY POINTS

- It is good design practice to partition the individual test cases from the main thread.
- It is good design practice to create an automated, self-checking testbench. This will save a significant amount of time down the road as the testbench grows.
- Initialize testbench clocks and resets with nonblocking assignments and update them with blocking assignments.
- Create test cases such that they can stand alone inside the main thread.
- Reference signals inside the design at the module boundaries whenever possible.
- MATLAB can be very useful when creating large or complex patterns for simulation.
- Code coverage checks the extent to which the design has been simulated and identifies any unsimulated structures.
- Gate-level simulations can be useful when estimating dynamic power dissipation.
- Timescale precision must be chosen to balance simulation accuracy against run-time.
- Intertial delays due to combinatorial logic should be modeled with continuous assignments.

Chapter 12

Coding for Synthesis

At the level of abstraction where logic is coded in an HDL language, synthesis optimizations can only take a designer so far when meeting design requirements. At its very fundamental level, a synthesis tool will follow the coding structure and map the logic according to the architecture laid out in the RTL. Only for very regular structures such as FSMs, RAMs, and so forth can a synthesis tool extract the functionality, identify alternative architectures, and implement accordingly.

Aside from optimization, a fundamental guiding principle when coding for synthesis is to minimize, if not eliminate, all structures and directives that could potentially create a mismatch between simulation and synthesis. A good coding style typically ensures that the RTL simulation will behave the same as the synthesized netlist. One class of deviations are the vendor-supplied directives that can be added to the RTL code in the form of special comments (that are ignored by the simulation tool) that will cause the synthesis tool to interpret a logic structure in a way that is not obvious from the RTL code itself.

During the course of this chapter, we will discuss the following topics:

- Creating efficient decision trees.
    Trade-offs between priority and parallel structures
    Dangers of the "parallel_case" and "full_case" directives
    Dangers of multiple control branches
- Coding style traps.
    Usage of blocking and nonblocking assignments
    Proper and improper usage of for-loops
    Inferrence of combinatorial loops and latches
- Design partitioning and organization.
    Organizing data path and control structures
    Modular design
- Parameterizing a design for reuse.

Advanced FPGA Design. By Steve Kilts
Copyright © 2007 John Wiley & Sons, Inc.

12.1 DECISION TREES

In the context of FPGA design, we refer to a decision tree as the sequence of conditions that are used to decide what action the logic will take. Usually, this breaks
down to if/else and case structures. Consider a very simple register write
example:

```
module regwrite(
    output reg   rout,
    input        clk,
    input [3:0]  in,
    input [3:0]  ctrl);

    always @(posedge clk)
        if(ctrl[0])      rout <= in[0];
        else if(ctrl[1]) rout <= in[1];
        else if(ctrl[2]) rout <= in[2];
        else if(ctrl[3]) rout <= in[3];
endmodule
```

This type of if/else structure can be conceptualized according to the mux structure shown in Figure 12.1.

This type of decision structure could be implemented in a number of different
ways depending on speed/area trade-offs and required priority. This section
describes how various decision trees can be coded and constrained to target different synthesized architectures.

12.1.1 Priority Versus Parallel

Inherent in the if/else structure is the concept of priority. Those conditions that
occur first in the if/else statement are given priority over others in the tree. A
higher priority with the structure above would correspond with the muxes near
the end of the chain and closer to the register.

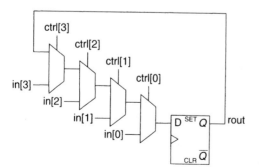

Figure 12.1 Simple priority with serialized mux structure.

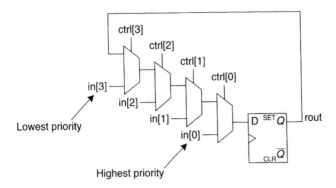

Figure 12.2   Priority placement.

In Figure 12.2, if bit 0 of the control word is set, in[0] will be registered regardless of the state of the other bits of the control word. If bit 0 of the control word is not set, then the states of the other bits are used to determine the signal that is passed to the register. In general, a bit will only be used to select the output if all bits ahead of it (in this case the LSBs) are not set. This is true of the priority mux implementation shown in Figure 12.3.

Regardless of the final implementation of the if/else structure, a higher priority is given to the conditional statements that occur previous to any given condition.

If/else structures should be used when the decision tree has a priority encoding.

Case structures, on the other hand, are often (but not always) used in circumstances where all conditions are mutually exclusive. In other words, they can be used to optimize the decision tree when only one condition can be true at any given time. For instance, when making a decision based on the value of some

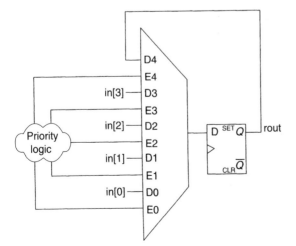

Figure 12.3   Priority mux.

other multibit net or register (say for an address decoder), only one condition can be true at one time. This is true of the decode operation that is implemented above with an if/else structure. To implement the exact same functionality in Verilog, a case statement can be used:

```
case(1)
    ctrl[0]: rout <= in[0];
    ctrl[1]: rout <= in[1];
    ctrl[2]: rout <= in[2];
    ctrl[3]: rout <= in[3];
endcase
```

Due to the fact that the case statement is available as an alternative to the if/else structure, many novice designers assume that this will implement a priority-less decision tree automatically. This happens to be true for the more rigorous language VHDL but is not the case for Verilog as can be seen in the implementation of the case statement in Figure 12.4.

As can be seen in Figure 12.4, the default is such that the priorities are encoded to set the appropriate enable pins on the mux. This leads many designers into a trap. To remove the priority encoding, it is possible to use the synthesis

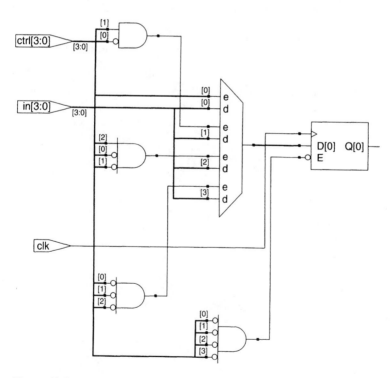

Figure 12.4 Priority-encoded logic.

directive "parallel_case" to implement a truly parallel structure. The syntax shown below will work with Synplicity and XST synthesis tools.

```
// DANGEROUS CASE STATEMENT
case(1)  // synthesis parallel_case
```

This directive can typically be added to the synthesis constraints. If this directive is used, it is certainly better to add it to the constraints so it is not "hidden" in the code if the designer needs to port to a new tool. This directive informs the synthesis tool that the cases are mutually exclusive and that it may forego any priority encoding as shown in Figure 12.5.

Here, all inputs are selected based on enable bits that are assumed to be mutually exclusive. This implementation is faster and consumes less logic resources.

Note that the parallel_case directive is a synthesis-only directive, and thus mismatches between simulation and actual implementation can occur. Frequent use of synthesis directives in general is bad design practice. It is better to code the RTL such that both the synthesis and simulation tools recognize the parallel architecture.

Use of the parallel_case directive is generally bad design practice.

Good coding practice dictates that priority encoders should be implemented with if/else statements, and structures that are parallel by design should be coded with case statements. There is typically no good reason to use the parallel_case

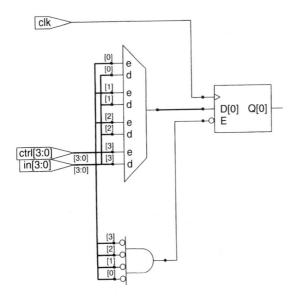

Figure 12.5   No priority encoding.

directive. Some designers can successfully use these statements to optimize one-hot decoders, but because of the risks involved it is better to never use these statements at all. If the synthesis tool reports that the case structure is not parallel, then the RTL must be changed to make it so. If it is truly a priority condition, then an if/else should be used in its place.

12.1.2 Full Conditions

In the decision trees examined thus far, if none of the conditions of a case statement were true, the synthesis tool has fed the output of the register back around to the decision tree as a default condition (this behavior will of course depend on the default implementation style of the synthesis tool, but for this section we will assume it is true). Even with the "full_case" implementation, there is logic that will disable the register if none of the selection bits are asserted. The assumption is that if no conditions are met, the value does not change.

One option available to the designer is to add a default condition. This may or may not be the current value, but it avoids the condition where the tool automatically latches the current value assuming that the output is assigned a value under each case condition. The register enable will be eliminated with this default condition as shown in the following modification to the case statement.

```
// DANGEROUS CASE STATEMENT
module regwrite(
   output reg   rout,
   input        clk,
   input [3:0] in,
   input [3:0] ctrl);

   always @(posedge clk)
     case(1) // synthesis parallel_case
       ctrl[0]: rout <= in[0];
       ctrl[1]: rout <= in[1];
       ctrl[2]: rout <= in[2];
       ctrl[3]: rout <= in[3];
       default: rout <= 0;
     endcase
endmodule
```

As can be seen in Figure 12.6, the default condition is now explicit and is implemented as an alternative input to the mux. Although the flip-flop no longer requires an enable, the total logic resources have not necessarily decreased. Also note that if not every condition defines an output for the register (this often occurs when multiple outputs are assigned within a single case statement), neither a default condition nor any synthesis tag will prevent the creation of a latch. To ensure that a value is always assigned to the register, an initial assignment can be used to assign a value to the register prior to the case statement. This is shown in the following example.

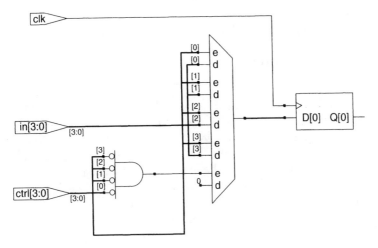

Figure 12.6 Encoding for default condition.

```
module regwrite(
    output reg   rout,
    input        clk,
    input  [3:0] in,
    input  [3:0] ctrl);

    always @(posedge clk) begin
        rout <= 0;
        case(1)
            ctrl[0]: rout <= in[0];
            ctrl[1]: rout <= in[1];
            ctrl[2]: rout <= in[2];
            ctrl[3]: rout <= in[3];
        endcase
    end
endmodule
```

This type of coding style eliminates the need for a default case and also ensures that the register is assigned to the default value if no other assignment is defined.

A synthesis directive similar to the parallel_case statement is the full_case directive. This directive informs the synthesis tool that all cases have been covered and that an implied default condition is not required. In general, the full_case directive is dangerous and can create a number of traps leading to incorrect or inefficient synthesis as well as mismatches with simulation.

Use of the full_case directive is generally bad design practice.

Full conditions can be designed with proper coding styles that completely avoid this directive as shown in the following example.

The full_case directive can be added in a similar way to the parallel_case directive as shown in the following example.

```
// DANGEROUS CASE STATEMENT
case(1) // synthesis full_case
```

Figure 12.7 illustrates the implementation with this directive.

The full_case statement tells the synthesis tool that all possible conditions have been covered by the case statement regardless of how the tool interprets the conditions. This implies that a default condition such as holding its current value is not necessary. As can be seen from the above implementation, all logic for maintaining the current value has been removed. All cases are assumed to be covered, and thus the only logic remaining is the mux itself.

The full_case directive, like the parallel_case directive, is synthesis-only meaning that it will be ignored by simulation. This makes the full_case directive dangerous in that mismatches between simulation and synthesis may occur. Specifically, if an output value is not provided by each condition, the simulation tool will latch the current value, whereas the synthesis tool will consider this a "don't care."

> Parallel_case and full_case can cause mismatches between simulation and synthesis.

The recommended approach is to avoid this constraint and to guarantee full coverage by design; that is, by using a default condition and setting default values prior to the case statement as shown above. This will make the code more portable and reduce the possibility of undesirable mismatches.

One of the biggest dangers when setting FPGA synthesis options is the allowance of a default setting whereby all case statements are automatically assumed to be full_case, parallel_case, or both. It is quite frankly shocking that any vendors actually provide this as an option. In practice, this option should never be used. This type of option only creates hidden dangers in the form of improperly

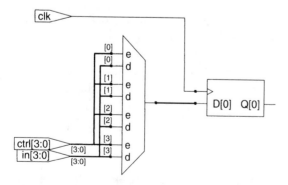

Figure 12.7   No default condition.

synthesized code that may not be discovered with basic in-system testing and certainly not in simulation.

12.1.3 Multiple Control Branches

One common mistake (in the form of a poor coding style) is to disconnect the control branches for a single register. In the following example, oDat is assigned to two different values in two unique decision trees.

```
// BAD CODING STYLE
module separated(
    output reg oDat,
    input      iClk,
    input      iDat1, iDat2, iCtrl1, iCtrl2);

    always @(posedge iClk) begin
        if(iCtrl2) oDat <= iDat2;
        if(iCtrl1) oDat <= iDat1;
    end
endmodule
```

Because there is no way to tell if iCtrl1 and iCtrl2 are mutually exclusive, this coding is ambiguous, and the synthesis tool must make certain assumptions for the implementation. Specifically, there is no explicit way to handle the priority when both conditions are true simultaneously. Thus, the synthesis tool must assign a priority based on the sequence in which these conditions occur. In this case, if the condition appears last, it will take priority over the first.

Based on Figure 12.8, iCtrl1 has priority over iCtrl2. If the order of these is swapped, the priority will likewise be swapped. This is the opposite behavior of an if/else structure that will give priority to the first condition.

It is good design practice to keep all register assignments inside one single control structure.

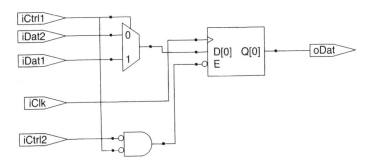

Figure 12.8 Implementation with implicit priority.

12.2 TRAPS

Because of the fact that behavioral HDL is very flexible in terms of its ability to describe functionality relative to the constrictive nature of synthesizable RTL, there will naturally be a number of traps that a designer can fall into when they do not understand how various structures are interpreted by synthesis tools. This section identifies a number of traps and discusses design methods to avoid them.

12.2.1 Blocking Versus Nonblocking

In the world of software design, functionality is created by defining operations that are executed in a predefined sequence. In the world of HDL design, this type of execution can be thought of as *blocking*. This means that future operations are blocked (they are not executed) until after the current operation has completed. All future operations are under the assumption that all previous operations have completed and all variables in memory have been updated. A *nonblocking* operation executes independent of order. Updates are triggered off of specified events, and all updates occur simultaneously when the trigger event occurs.

HDL languages such as Verilog and VHDL provide constructs for both blocking and nonblocking assignments. Failure to understand where and how to use these can lead not only to unexpected behavior but also to mismatches between simulation and synthesis. For example, consider the following code.

```
module blockingnonblocking(
   output reg out,
   input      clk,
   input      in1, in2, in3);
   reg        logicfun;

   always @(posedge clk) begin
   logicfun <= in1 & in2;
   out      <= logicfun | in3;
   end
endmodule
```

The logic is implemented as a logic designer would expect as shown in Figure 12.9.

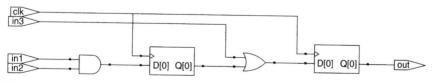

Figure 12.9 Simple logic with nonblocking assignments.

In the implementation shown in Figure 12.9, both the signals "logicfun" and "out" are flip-flops, and any changes on "in1" or "in2" will take two clock cycles to propagate to "out." A move to blocking assignments appears to be only a subtle change.

```
// BAD CODING STYLE
logicfun = in1 & in2;
out      = logicfun | in3;
```

In the above modification, the nonblocking statements have been changed to blocking. This means that out will not be updated until logicfun has been updated, and both updates must occur on one event of the clock.

As can be seen from Figure 12.10, by changing the assignments to blocking, we have effectively eliminated the register for logicfun and changed the timing through the entire design. This is certainly not to say that the same functionality cannot be accomplished with blocking assignments. Consider the following modification.

```
// BAD CODING STYLE
out      = logicfun | in3;
logicfun = in1 & in2;
```

In the above modification, we force the out register to be updated before logicfun, which forces a 2-clock cycle delay for the inputs in1 and in2 to propagate to out. This will give us the intended logic implementation, but with a less straightforward approach. In fact, for many logic structures with a significant amount of complexity, this is not a clean or even a feasible approach. One temptation may be to use independent always statements for each assignment.

```
// BAD CODING STYLE
always @(posedge clk)
  logicfun = in1 & in2;

always @(posedge clk)
  out      = logicfun | in3;
```

Despite the fact that these assignments are split into seemingly parallel blocks, they will not simulate as such. This type of coding style should be avoided.

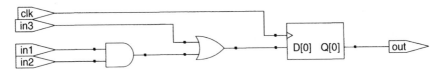

Figure 12.10   Incorrect implementation with blocking assignments.

One case where blocking assignments often arise is with operations that require a relatively large number of default conditions. In the following code example that uses nonblocking assignments, the control signal ctrl defines which input is assigned to the corresponding output. The remaining outputs are assigned to zero.

```
// POOR CODING STYLE
module blockingnonblocking(
  output reg [3:0] out,
  input            clk,
  input      [3:0] ctrl, in);

  always @(posedge clk)
    if(ctrl[0]) begin
      out[0]   <= in[0];
      out[3:1] <= 0;
    end
    else if(ctrl[1]) begin
      out[1]   <= in[1];
      out[3:2] <= 0;
      out[0]   <= 0;
    end
    else if(ctrl[2]) begin
      out[2]   <= in[2];
      out[3]   <= 0;
      out[1:0] <= 0;
    end
    else if(ctrl[3]) begin
      out[3]   <= in[3];
      out[2:0] <= 0;
    end
    else
      out      <= 0;
endmodule
```

In each decision branch in the above implementation, all outputs that are not assigned must be set to zero. Each branch contains a single output assignment to an input and three zero assign statements. To simplify the code, blocking statements are sometimes used with initial assignments as shown in the following example.

```
// BAD CODING STYLE
module blockingnonblocking(
  output reg [3:0] out,
  input            clk,
  input      [3:0] ctrl, in);

  always @(posedge clk) begin
    out            = 0;
```

```
    if(ctrl[0])        out[0] = in[0];
    else if(ctrl[1]) out[1] = in[1];
    else if(ctrl[2]) out[2] = in[2];
    else if(ctrl[3]) out[3] = in[3];
  end
endmodule
```

Because the last assignment is the one that "sticks," the above modification sets an initial value for all output bits and then only changes one output as necessary. Although this code will synthesize to the same logic structure as the more complex nonblocking structure, race conditions may appear in simulation. Although it is less intuitive, nonblocking assignments can be used to accomplish the same thing with a similar coding style as shown below.

```
module blockingnonblocking(
    output reg  [3:0] out,
    input             clk,
    input       [3:0] ctrl, in);

  always @(posedge clk) begin
    out                   <= 0;
    if(ctrl[0])        out[0] <= in[0];
    else if(ctrl[1]) out[1] <= in[1];
    else if(ctrl[2]) out[2] <= in[2];
    else if(ctrl[3]) out[3] <= in[3];
  end
endmodule
```

This coding style is superior as the race conditions have been eliminated with the nonblocking assignments. There are a number of widely accepted guidelines regarding blocking and nonblocking assignments when coding for synthesis:

Use blocking assignments to model combinatorial logic.

Use nonblocking assignments to model sequential logic.

Never mix blocking and nonblocking assignments in one always block.

Violating these guidelines will likely lead to mismatches in simulation versus synthesis, poor readability, decreased simulation performance, and hardware errors that are difficult to debug.

12.2.2 For-Loops

C-like looping structures such as the for-loop can present a trap to a designer with a background in software design. The reason for this is, unlike the C software language, these loops cannot typically be used for algorithmic iterations in synthesizable code. Instead, HDL designers will typically use these looping structures to minimize typing a large array of similar assignments that operate on similar elements. For instance, a software designer may use a for-loop to take X to the power of N as shown in the following snippet.

```
PowerX = 1;
for(i=0;i<N;i++) PowerX = PowerX * X;
```

This algorithmic loop uses iteration to perform a multiply operation N times. Each time through the loop, the running variable is updated. This works well in software because for every loop iteration, an internal register is updated with the current value of PowerX.

Synthesizable HDL, in contrast, does not have any implied registering that occurs during an iterative loop. Instead, all register operations are defined explicitly. If a designer attempted to create the above structure in a similar way with synthesizable HDL, they might end up with something that looks like the following code segment.

```
// BAD CODING STYLE
module forloop(
   output reg [7:0] PowerX,
   input        [7:0] X, N);
   integer          i;

   always @* begin
     PowerX = 1;
     for(i=0;i<N;i=i+1)
       PowerX = PowerX * X;
   end
endmodule
```

This will work in a behavioral simulation and, depending on the synthesis tool, may be synthesizable to gates. XST will not synthesize this code without a fixed value of N, whereas Synplify will synthesize this loop based on the worst-case value of N. The end result if this is indeed synthesized will be a loop that is completely unrolled into a massive block of logic that runs extremely slow. A design that manages the registers during each iteration of the loop may utilize control signals as shown in the following example.

```
module forloop(
   output reg [7:0] PowerX,
   output reg       Done,
   input           Clk, Start,
   input     [7:0] X, N);
   integer          i;

   always @(posedge Clk)
     if(Start) begin
       PowerX <= 1;
       i      <= 0;
```

```
    Done      <= 0;
end
else if(i  < N) begin
    PowerX <= PowerX * X;
    i         <= i + 1;
    end
else
    Done      <= 1;
endmodule
```

In the above design, the power function will be an order of magnitude smaller and will run an order of magnitude faster than the "software-like" implementation.

For-loops should not be used to implement software-like iterative algorithms.

In contrast with the previous example, understanding the proper use of for-loops can help to create readable and efficient HDL code. As mentioned earlier, for-loops are often used as a short form to reduce the length of repetitive but parallel code segments. For instance, the following code generates an output by taking every bit in X and applying the XOR operation with every even bit of Y.

```
Out[0]   <= Y[0]   ^ X[0];
Out[1]   <= Y[2]   ^ X[1];
Out[2]   <= Y[4]   ^ X[2];
...
Out[31] <= Y[62] ^ X[31];
```

To write this out in long form would require 32 lines and a fair amount of typing. To condense this, and to make it more readable, a for-loop can be used to replicate the operation for each bit.

```
always @(posedge Clk)
    for(i=0;i<32;i=i+1) Out[i] = Y[i*2] ^ X[i];
```

As can be seen from the above example, there are no feedback mechanisms in the loop. Rather, the for-loop is used to condense similar operations.

12.2.3 Combinatorial Loops

Combinatorial loops are logic structures that contain feedback without any intermediate synchronous elements. As can be seen from Figure 12.11, a combinatorial loop occurs when the output of a cloud of combinatorial logic feeds back to itself with no intermediate registers. This type of behavior is rarely desirable and typically indicates an error in the design or implementation. In Chapter 18, we will discuss how to handle the timing analysis of such a structure, but for this

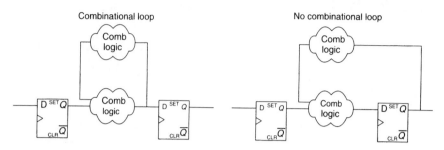

Figure 12.11   Combinatorial versus sequential loops.

discussion we will discuss traps that may create such a structure and how to avoid them. Consider the following code segment.

```
// BAD CODING STYLE
module combfeedback(
    output out,
    input a);
    reg b;

    // BAD CODING STYLE: this will feed b back to b
    assign out = b;

    // BAD CODING STYLE: incomplete sensitivity list
    always @(a)
        b = out ^ a;
endmodule
```

The above module represents a behavioral description that in simulation may behave as follows: when the wire "a" changes, the output is assigned the result of the current output XOR "a." The output only changes when "a" changes and does not exhibit any feedback or oscillatory behavior. In FPGA synthesis, however, an always structure describes the behavior of either registered or combinatorial logic. In this case, the synthesis tool will likely expand the sensitivity list (currently containing only "a") to include all inputs assuming the structure is combinatorial. When this happens, the feedback loop is closed and will be implemented as an XOR gate that feeds back on itself. This is represented in Figure 12.12.

This type of structure is very problematic as it will oscillate any time the input "a" is a logic-1. The Verilog listed above describes a circuit with a very poor coding style. The designer clearly did not have the hardware in mind and

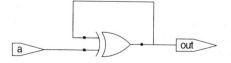

Figure 12.12   Accidental combinatorial feedback.

will see a drastic mismatch between simulation and synthesis. As a matter of good coding practice, all combinatorial structures should be coded such that all inputs to the expressions contained within the always block are listed in the sensitivity list. If this had been done in the prior example, the problem would have been detected prior to synthesis.

12.2.4 Inferred Latches

Special types of combinatorial feedback can actually infer sequential elements. The following module models a latch in the typical manner.

```
// latch inferrence
module latch (
    input       iClk, iDat,
    output reg oDat);

    always @*
        if(iClk) oDat <= iDat;
endmodule
```

Whenever the control is asserted, the input is passed directly to the output. When the control is deasserted, the latch is disabled. A very common coding mistake is to create a combinatorial if/else tree and forget to define the output for every condition. The implementation will contain a latch and will usually indicate a coding error.

Latches are typically not recommended for FPGA designs, but it is possible to design and perform timing analysis with these devices (timing analysis with latches is discussed in future chapters). Note that there are other ways of accidentally inferring latches and are more than likely unintended. In the following assignment, the default condition is the signal itself.

```
// BAD CODING STYLE
assign O = C ? I: O;
```

Instead of inferring a mux with feedback to one of its inputs (which would not be desirable anyway), some synthesis tools will infer a latch that enables passthrough of the input whenever the control is asserted. The problem with this is that a timing end point (the latch) is inserted into a path that was most likely not designed to have an intermediate sequential element. This type of latch inference typically indicates an error in the HDL description.

Functions

Latches are typically not recommended for FPGA designs and can very easily become implemented improperly or not implemented at all. One example is through the use of a function call. Consider, for example the typical instantiation of a latch encapsulated into a function as shown in the following example.

```
// BAD CODING STYLE
module latch (
    input       iClk, iDat,
    output reg oDat);

    always @*
        oDat <= MyLatch(iDat, iClk);

    function MyLatch;
    input D, G;
        if(G) MyLatch = D;
    endfunction
endmodule
```

In this case, the conditional assignment of the input to the output is pushed into a function. Despite the seemingly accurate representation of the latch, the function will always evaluate to combinatorial logic and will pass the input straight through to the output.

12.3 DESIGN ORGANIZATION

Anyone who has worked with a team of engineers to design a large FPGA understands the importance of organizing a design into useful functional boundaries and designing for reusability and expandability. The goal when organizing a design at the top level is to create a design that is easier to manage on a module-by-module basis, to create readable and reusable code, and to create a basis that will allow the design to scale. This section discusses some of the issues to consider when architecting a design that will affect the readability, reusability, and synthesis efficiency.

12.3.1 Partitioning

Partitioning refers to the organization of the design in terms of modules, hierarchy, and other functional boundaries. The partitioning of a design should be considered up front, as major changes to the design organization will become more difficult and expensive as the project progresses. Designers can easily wrap their minds around one piece of functionality, and this will allow them to design, simulate, and debug their block in an efficient manner.

12.3.1.1 Data Path Versus Control

Many architectures can be partitioned into what is called data path and control structures. The data path is typically the "pipe" that carries the data from the input of the design to the output and performs the necessary operations on the data. The control structure is usually one that does not carry or process the data

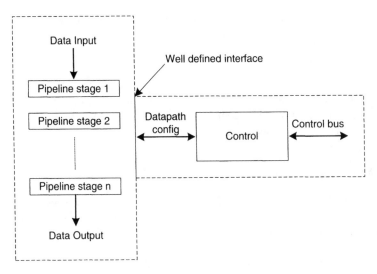

Figure 12.13   Partitioning between datapath and control.

through the design but rather configures the data path for various operations. Figure 12.13 illustrates a logic partitioning between the data path and the control.

With the partitioning shown in Figure 12.13, it would be logical to place the data path and control structures into different modules and to clearly define the interface for the individual designers. This makes it easier not only for various logic designers to divide up the design activities but also for the optimizations that may be required downstream.

Data path and control blocks should be partitioned into different modules.

Because the data path is often the critical path of the design (the throughput of the design will be related to the timing of the pipeline), it may be required that a floorplan is designed for this path to achieve maximum performance. The control logic, on the other hand, will often have slower timing requirements placed on it because it is not a part of the primary data path.

For instance, it is very common to have a simple SPI (Serial Peripheral Interface) or I^{2}C (Inter-IC) type bus to set up control registers in the design. If the pipeline is running at perhaps hundreds of megahertz, there will certainly be a large discrepancy between the two timing requirements. Thus, if a floorplan is required for the data path, the control logic can usually remain unconstrained (spatially) and scattered around the pipeline wherever it fits and as decided by the automatic place and route tool.

12.3.1.2 Clock and Reset Structures

Good design practice dictates that any given module should have only one type of clock and one type of reset. If, as is the case in many designs, there are multiple clock domains and/or reset structures, it is important to partition the hierarchy so that they are separated by different modules.

It is good design practice to use only one clock and only one type of reset in each module.

Other chapters discuss the hazards involved with mixing clock and reset types in procedural descriptions, but if any given module has only one clock and reset, these problems are less likely to arise.

12.3.1.3   Multiple Instantiations

If there are cases where certain logic operations occur more than once in a particular module (or across multiple modules), a natural partition to the design would be to group that block into a separate module and push it into the hierarchy for multiple instantiations.

There are a number of advantages to the partitioning described in Figure 12.14. First of all, it will be much easier to assign blocks of functionality to designers independent of one another. One designer can focus on the top-level design, organization, and simulation, while another designer can focus on the functionality specific to the submodule. If the interfaces are clearly defined, this type of group design can work very well. If, however, both designers are developing within the same module, greater confusion and difficulty can occur. Additionally, the submodules can be reused in other areas of the design or perhaps in different designs altogether. It is typically much easier to reinstantiate an existing module rather than cut and paste out of larger modules and redesign the corresponding interfaces.

One difficulty that may arise with such a strategy is the case where there are slight variations on the individual modules such as data width, iteration count, and so forth. These cases are addressed with a design method called parameterization, whereby like-kind modules can share a common code-base that is parameterizable on an instance-by-instance basis. The next section discusses this in more detail.

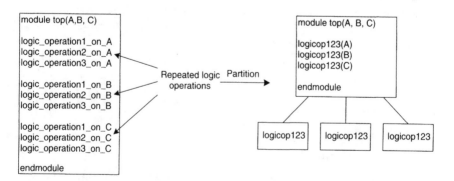

Figure 12.14   Modular design.

12.3.2 Parameterization

In the context of FPGA design, a parameter is a property of a module that can be changed either on a global sense or on an instance-by-instance basis while maintaining the root functionality of the module. This section describes the forms of parameterization and how they can be leveraged for efficient coding for synthesis.

12.3.2.1 Definitions

Parameters and definitions are similar and in many cases can be used interchangeably. However, there are a number of circumstances where one is preferred over another for efficient, readable, and modular design. Definitions are typically used either to define global values that are constant across all modules or to provide compiler time directives for inclusion and exclusion of portions of code. In Verilog, definitions are utilized with the 'define statement, and for compiler time controls with subsequent 'ifdef statements. Examples of global definitions may be to define design-wide constants such as:

```
'define CHIPID 8'hC9 // global chip ID
'define onems 90000 // approximately 1ms with an 11ns
                       clock
'define ulimit16 65535 // upper limit of an unsigned
                         16-bit word
```

The definitions listed above are examples of global "truisms" that will not change from submodule to submodule. The other use for a global define is to specify compile-time directives for code selection. One very common application is that of ASIC prototyping in an FPGA. There will often be slight modifications to the design (particularly at the I/O and global structures) that will be different between the ASIC and FPGA. For example, consider the following lines in a defines file:

```
'define FPGA
//'define ASIC
```

In the top-level module, there may be entries such as:

```
'ifdef ASIC
input TESTMODE;
output TESTOUT;
'endif

'ifdef FPGA
output DEBUGOUT;
'endif
```

In the above code example, test pins must be included for ASIC test insertion but have no meaning in the FPGA implementation. Thus, a designer would only include these placeholders in ASIC synthesis. Likewise, the designer may have outputs that are used for debug in the FPGA prototype but will not be included in the final ASIC implementation. Global definitions allow the designer to maintain a single code base with the variations included in-line.

Ifdef directives should be used for global definitions.

To ensure that definitions are applied in a global sense and do not contradict one another, it is recommended that a global definitions file be created that can be included in all design modules. Thus, any global parameters that change can be modified in one central location.

12.3.2.2 Parameters

Unlike global definitions, parameters are typically localized to specific modules and can vary from instantiation to instantiation. A very common parameter is that of size or bus width as shown in the following example of a register.

```
module paramreg #(parameter WIDTH = 8) (
    output reg [WIDTH-1:0] rout,
    input                  clk,
    input      [WIDTH-1:0] rin,
    input                  rst);

    always @(posedge clk)
        if(!rst) rout <= 0;
        else     rout <= rin;
endmodule
```

The above code example illustrates a simple parameterized register with variable width. Although the parameter defaults to 8 bits, every individual instantiation can modify the width for that instantiation only. For instance, a module at a higher level in the hierarchy could instantiate the following 2-bit register:

```
// CORRECT, BUT OUTDATED PARAMETER PASSING
paramreg #(2) r1(.clk(clk), .rin(rin), .rst(rst),
.rout(rout));
```

Or the following 22-bit register:

```
// CORRECT, BUT OUTDATED PARAMETER PASSING
paramreg #(22) r2(.clk(clk), .rin(rin), .rst(rst),
.rout(rout));
```

As can be seen from the above instantiations, the same code base for "paramreg" was used to instantiate two registers with different properties. Also note that the base functionality of the module did not change between the instantiations (a register) but only a specific property of that functionality (size).

Parameters should be used for local definitions that will change from module to module.

Parameterized code such as this is useful when different modules of similar functionality but slightly different characteristics are required. Without parameterization, the designer would need to maintain a large code base for variations of the same module where changes would be tedious and error-prone. The alternative would be to use the same module across instantiations for which the characteristics are not optimal.

An alternative to the above parameter definition is the use of the "defparam" command in Verilog. This allows the designer to specify any parameter in the design hierarchy. The danger here is that because parameters are typically used at specific module instances and are not seen outside of that particular instance (analogous to local variables in software design), it is easy to confuse the synthesis tool and create mismatches to simulation. A common scenario that illustrates poor design practice is shown in Figure 12.15.

Figure 12.15 illustrates the practice of passing parameters through the hierarchy from top to bottom through instantiations but then redefining the top-level parameter from a submodule in the hierarchy. This is poor design practice not only from an organization and readability standpoint; there also exists the potential for a mismatch between simulation and synthesis. Although the simulation tool may simulate properly and as the designer intended, synthesis tools often evaluate parameters from the top down and construct the physical structure accordingly. It is therefore recommended that if defparams are used, they should always be included at the module instantiation corresponding with the parameter it is defining.

A superior form of parameterization was introduced in Verilog-2001, and this is discussed in the next section.

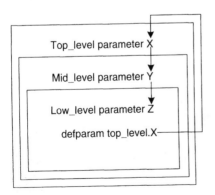

Figure 12.15   Abuse of defparam.

12.3.2.3 Parameters in Verilog-2001

An improved method of parameterization was introduced in Verilog-2001. In older versions of Verilog, the passing of parameter values was either cryptic or hard to read with positional parameter passing, and a number of dangers arose with defparams as discussed in the previous section. Ideally, a designer would pass a list of parameter values to a module in a similar way that signals are passed between the I/O of a module. In Verilog-2001, the parameters can be referenced by name outside of the module, eliminating the readability issues as well as the dangers associated with defparam. For instance, the instantiation of paramreg from the previous section would be modified to include the parameter name.

```
paramreg #(.WIDTH(22)) r2(.clk(clk), .rin(rin), .rst(rst),
.rout(rout));
```

This unlocks the position requirement, enhances the readability of the code, and reduces the probability of human error. This type of named parameterization is highly recommended.

Named parameter passing is superior to positional parameter passing or the defparam statement.

The other major enhancement to parameterization in Verilog-2001 is the "localparam." The localparam is a Verilog parameter version of a local variable. The localparam can be derived from an expression using other parameters and is confined to the particular instantiation of the module in which it resides. For instance, consider the following parameterized multiplier.

```
// MIXED HEADER STYLE FOR LOCALPARAM
module multparam #(parameter WIDTH1 = 8, parameter WIDTH2 = 8)
                  (oDat, iDat1, iDat2);
    localparam      WIDTHOUT = WIDTH1 + WIDTH2;
    output [WIDTHOUT-1:0] oDat;
    input [WIDTH1-1:0]   iDat1;
    input [WIDTH2-1:0]   iDat2;


    assign oDat = iDat1 * iDat2;
endmodule
```

In the above example, the only parameters that need to be defined externally are the widths of the two inputs. Because the designer is always assuming that the width of the output is the sum of the widths of the inputs, this parameter can be derived off of the input parameters and not redundantly calculated externally. This makes the job of the designer easier and eliminates the possibility that the output size does not match the sum of the input sizes.

Currently, localparam is not supported in the module header, and thus the port list must be stated redundantly (Verilog-1995 style) if the localparam is used in the I/O list. Regardless, localparam is recommended whenever it can be derived off of other input parameters as it will further reduce the possibility of human error.

12.4 SUMMARY OF KEY POINTS

- If/else structures should be used when the decision tree has a priority encoding.
- Use of the parallel_case directive is generally bad design practice.
- Use of the full_case directive is generally bad design practice.
- Parallel_case and full_case can cause mismatches between simulation and synthesis.
- It is good design practice to keep all register assignments inside one single control structure.
- Use blocking assignments to model combinatorial logic.
- Use nonblocking assignments to model sequential logic.
- Never mix blocking and nonblocking assignments in one always block.
- For-loops should not be used to implement software-like iterative algorithms.
- Data path and control blocks should be partitioned into different modules.
- It is good design practice to use only one clock and only one type of reset in each module.
- Ifdef directives should be used for global definitions.
- Parameters should be used for local definitions that will change from module to module.
- Named parameter passing is superior to positional parameter passing or the defparam statement.

Chapter 13

Example Design: The Secure Hash Algorithm

The secure hash algorithm (SHA) defines a method for creating a condensed representation of a message (the message digest) that is computationally infeasible to create without the message itself. This property makes the SHA useful for applications such as digital signatures to verify the authenticity of a message or for more ancillary applications such as random number generation.

One of the advantages of the SHA algorithms defined by NIST is that they lend themselves to an easy implementation in hardware. All of the operations are relatively simple to code with logical operations that are efficiently implemented in FPGAs. The objective of this chapter is to implement the SHA-1 standard in a parameterized fashion and evaluate the effects of varying the parameters.

13.1 SHA-1 ARCHITECTURE

The various SHA standards are related to the hash size, which corresponds directly with the level of security (particularly for applications such as digital signatures). The SHA standard that is considered in this chapter is the most basic, the SHA-1. The SHA-1 algorithm operates on 32-bit words, and each intermediate calculation (hash) is computed from 512-bit blocks (16 words). The message digest is 160 bits, which is the condensed representation of the original message.

For the purposes of this illustration, we will assume that the initial message has been properly padded and parsed. The 160-bit hash is initialized with five 32-bit words defined by the SHA standard and are labeled $H_0^{(i)}$, $H_1^{(i)}$, $H_2^{(i)}$, $H_3^{(i)}$, and $H_4^{(i)}$ (with initialized values $H^{(0)}$). The message schedule is represented as 80 words W_0, W_1, ..., W_{79}, the five working variables are represented by registers A–E, and the one temporary word is represented as T. The basic architecture is shown in Figure 13.1.

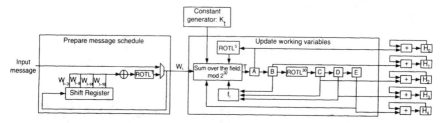

Figure 13.1 Basic SHA-1 architecture.

Table 13.1 Constant Generator Definition

K_t	Iteration t
5a827999	$0 \leq t \leq 19$
6ed9eba1	$20 \leq t \leq 39$
8f1bbcdc	$40 \leq t \leq 59$
ca62c1d6	$60 \leq t \leq 79$

Table 13.2 f_t Definition

f_t	Iteration t
(B & C) ∧ (~B & D)	$0 \leq t \leq 19$
B ∧ C ∧ D	$20 \leq t \leq 39$
(B & C) ∧ (C & D) ∧ (B & D)	$40 \leq t \leq 59$
B ∧ C ∧ D	$60 \leq t \leq 79$

Initially, the working variables are updated with the current hash values $H_0^{(i)}-H_4^{(i)}$ where the initial hash values themselves are defined as constants in the SHA specification. For a total of 80 loops, the message schedule must generate a unique W_t, which is added to a function of the working variables over the finite field mod 2^{32}. The constant generator is defined as shown in Table 13.1. The function f_t of B, C, and D is defined in Table 13.2. After 80 updates of the working variables, A–E are added to H_0-H_4, respectively, for the final hash value.

The implementation under consideration is a compact implementation designed to iteratively reuse logic resources. Both the message schedule and the working variable updates operate in an iterative manner as shown in the block diagram. It is not until the hash is complete that a new hash can begin. Note that because the previous hash must complete before the next can begin, there is very little in the design that can be pipelined. The code is shown below.

```
`define H0INIT 32'h67452301
`define H1INIT 32'hefcdab89
`define H2INIT 32'h98badcfe
```

```
`define H3INIT 32'h10325476
`define H4INIT 32'hc3d2e1f0

`define K0 32'h5a827999
`define K1 32'h6ed9eba1
`define K2 32'h8f1bbcdc
`define K3 32'hca62c1d6

module sha1 #(parameter WORDNUM = 16, parameter WORDSIZE = 32,
            parameter WSIZE = 480)(
output  [159:0]         oDat,
output reg              oReady,
input   [WORDSIZE-1:0]  iDat,
input                   iClk,
input                   iInitial, iValid);
reg     [6:0]           loop;
reg     [WORDSIZE-1:0]  H0, H1, H2, H3, H4;
reg     [WSIZE-1:0]     W;
reg     [WORDSIZE-1:0]  Wt, Kt;
reg     [WORDSIZE-1:0]  A, B, C, D, E;

// Hash functions
wire    [WORDSIZE-1:0]  f1,f2,f3, WtRaw, WtROTL1;
wire    [WORDSIZE-1:0]  ft;
wire    [WORDSIZE-1:0]  T;
wire    [WORDSIZE-1:0]  ROTLB; // rotate B left

// define SHA-1 function based on loop iteration
assign f1       = (B & C) ^ (~B & D);
assign f2       = B ^ C ^ D;
assign f3       = (B & C) ^ (C & D) ^ (B & D);
assign ft       = (loop < 21) ? f1 : (loop < 41) ? f2 : (loop < 61) ?
                  f3 : f2;

// Raw Wt computation before ROTL1
assign WtRaw    = {W[(WORDNUM-2)*WORDSIZE-1:(WORDNUM-3)*WORDSIZE]^
 W[(WORDNUM-7) * WORDSIZE-1:(WORDNUM-8) * WORDSIZE] ^
 W[(WORDNUM-13)* WORDSIZE-1:(WORDNUM-14)* WORDSIZE] ^
 W[(WORDNUM-15)* WORDSIZE-1:(WORDNUM-16)* WORDSIZE]};
// Wt ROTL by 1
assign WtROTL1 = {WtRaw[WORDSIZE-2:0],
 WtRaw[WORDSIZE-1]};

assign T        ={A[WORDSIZE-6:0],A[WORDSIZE-1:WORDSIZE-5]} +
                 ft + E + Kt + Wt;

assign ROTLB    = {B[1:0],B[WORDSIZE-1:2]};

assign oDat     = {H0, H1, H2, H3, H4};

// define Kt based on loop iteration
always @ (posedge iClk)
   if (loop < 20)      Kt <= `K0;
   else if (loop < 40) Kt <= `K1;
   else if (loop < 60) Kt <= `K2;
   else                Kt <= `K3;
```

```verilog
// message schedule
always @(posedge iClk) begin
  // preparing message schedule
  if(loop < WORDNUM)  Wt <= iDat;
  else                Wt <= WtROTL1;

  // shift iDat into MS position
  if((loop < WORDNUM-1) & iValid)
    W[WSIZE-1:0]          <= {iDat, W[WSIZE-1:WORDSIZE]};
  // shift Wt into MS position
  else if(loop > WORDNUM-1)
    W[WSIZE-1:0]          <= {Wt,W[(WORDNUM-1)*WORDSIZE-1:WORDSIZE]};
end

always @(posedge iClk)
  if(loop == 0) begin
    if(iValid) begin
      // initialize working variables
      if(!iInitial) begin
        A    <= 'H0INIT;
        B    <= 'H1INIT;
        C    <= 'H2INIT;
        D    <= 'H3INIT;
        E    <= 'H4INIT;

        H0   <= 'H0INIT;
        H1   <= 'H1INIT;
        H2   <= 'H2INIT;
        H3   <= 'H3INIT;
        H4   <= 'H4INIT;
      end
      else begin
        A    <= H0;
        B    <= H1;
        C    <= H2;
        D    <= H3;
        E    <= H4;
      end
      oReady <= 0;
      loop   <= loop + 1;
    end
    else
      oReady    <= 1;
  end
  else if(loop == 80) begin
    // compute intermediate hash
    H0        <= T + H0;
    H1        <= A + H1;
    H2        <= ROTLB + H2;
    H3        <= C + H3;
    H4        <= D + H4;
    oReady <= 1;
    loop   <= 0;
  end
```

```
else if(loop < 80) begin
  E    <= D;
  D    <= C;
  C    <= ROTLB;
  B    <= A;
  A    <= T;
  loop <= loop + 1;
end
else
  loop <= 0;
endmodule
```

Both the function $f_t(B,C,D)$ and K_t are implemented as muxes with the appropriate outputs selected based on the current iteration. The function $f_t(B,C,D)$ selects the appropriate transformation as shown in Figure 13.2.

Likewise, the constant generator simply selects predefined constants depending on the current iteration (Fig. 13.3).

Note that because only constants are selected, the synthesis tool was able to optimize certain bits. For the finite field addition, simple adders are used as shown in Figure 13.4.

Additions over mod 2^{32} are trivial in hardware because the modulus is handled automatically with a 32-bit register width containing the sum. Thus, the math for this type of finite field is easier than standard arithmetic because no overflow checks are necessary.

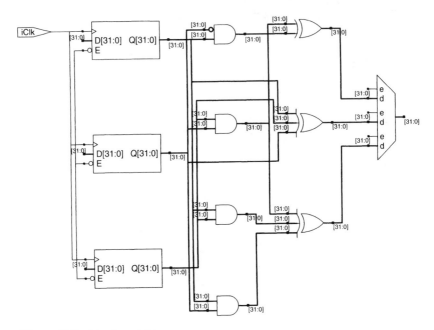

Figure 13.2 f_t implementation.

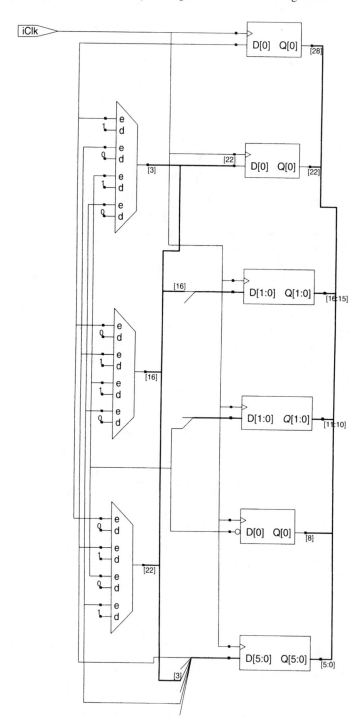

Figure 13.3   Constant generator implementation.

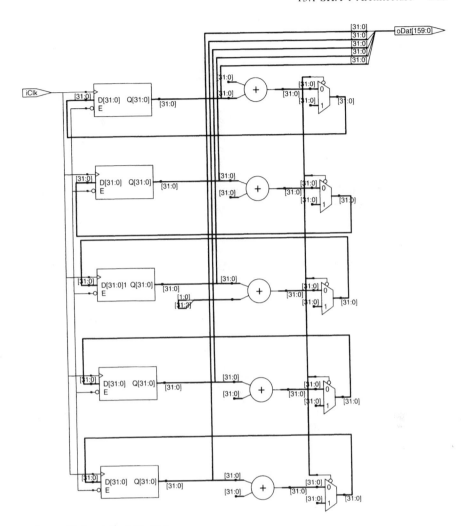

Figure 13.4 Finite field addition.

Note from the example that both definitions and parameters were used. Normally, the definitions would be contained in a separate defines.v file, but for this illustration they are included above the module. Note that the definitions contain the global or system-wide constants. The SHA specification defines the initial hash values and the constant table, both of which will never change. The parameters are used for module-specific parameters. The SHA specification defines the word size to be 32 bits and the block size to be 512 bits (16 words), but these are listed as parameters to serve as an illustration. It is feasible that a bus is driving data to the SHA core that is less than 32 bits, in which case parameters such as these as well as a parameterizable interface would be required. Note how all signal declarations and bit operations are based off of the basic parameters.

Table 13.3 Speed/Area Statistics Targeting a Xilinx Spartan-3

WORDNUM	WORDSIZE	Max clock frequency (MHz)	Area (Xilinx LUTs)
16	32	86	858
32	32	86	860
16	64	78	1728

Using this method, data paths of different widths can be created simply by changing the corresponding parameter, and the corresponding implementation results are described next.

13.2 IMPLEMENTATION RESULTS

In this section, we discuss speed/area trade-offs of parameter variances based off of a Xilinx Spartan-3 implementation. All variances were performed with the same code base with only the parameter definitions as independent variables. Table 13.3 shows the effects of varying both the width of the data path (WORDSIZE) and the number of input words (WORDNUM).

The implementation results in Table 13.3 illustrate the vast difference small parameterizations can make. The first parameter, WORDNUM, made very little difference as it simply adjusts the number of inputs that are initially muxed into the message schedule. This should add very little overhead as it will only impact a comparison and multiplex operation. The second parameter WORDSIZE, on the other hand, directly impacts the data path of the entire design. In this case, doubling the word size will effectively double the total number of resources required to implement this design.

Chapter 14

Synthesis Optimization

Most implementation tools for FPGA synthesis provide the designer with dozens of optimization options. The main problem most designers run into is that it is not clear what these options do exactly and more importantly how these can be used to actually optimize a design. Most designers never fully understand these optimization options, and after spending hours, days, or weeks playing around with the endless combinations, they find a formula that seems to give them the best results. After having gone down this road, few designers ever approach these optimizations beyond the formulas that have worked for them in the past. Thus, most optimizations go unused due to a fundamental lack of understanding and the difficulty in developing a full arsenal of heuristics.

This chapter describes the most important aspects of synthesis optimization at the implementation level based on tried-and-true real-world experience and will provide practical heuristics along the way that can be immediately leveraged by the reader. During the course of this chapter, we will discuss the following topics in detail:

- Trade-offs with speed versus area.
- Resource sharing for area optimization.
- Pipelining, retiming, and register balancing for performance optimization.
    The effect of reset on register balancing
    Handling resynchronization registers
- Optimizing FSMs.
- Handling black boxes.
- Physical synthesis for performance.

Advanced FPGA Design. By Steve Kilts
Copyright © 2007 John Wiley & Sons, Inc.

14.1 SPEED VERSUS AREA

Most synthesis tools provide switches that allow the designer to target speed versus area optimization. This seems like a no-brainer: If you want it to run faster, choose speed. If you want it to be smaller, choose area. This switch is misleading because it is a generalization of certain algorithms that can sometimes produce the opposite result (i.e., the design becomes slower after telling it to go faster). Before we understand why this happens, we must first understand what speed and area optimizations actually do to our design.

At the synthesis level, speed and area optimizations determine the logic topology that will be used to implement our RTL. At this level of abstraction, there is little known about the physical nature of the FPGA. Specific to this discussion, this would relate to the interconnect delay based on the place and route. Synthesis tools use what are called wire load models, which are statistical estimates of interconnect delay based on various criteria of the design. In an ASIC, this is accessible to the designer, but with FPGA design this is hidden behind the scenes. This is where the synthesis tool comes up with its estimates, which are often significantly different from the end result. Due to this lack of knowledge from the back end, synthesis tools will primarily execute gate-level optimizations. In high-end FPGA design tools, there exists a flow called *placement-based synthesis* to help close this loop, and this is discussed at the end of the chapter.

The synthesis-based gate-level optimizations will include things like state-machine encoding, parallel versus staggered muxing, logic duplication, and so on. As a general rule of thumb (although certainly not always true), faster circuits require more parallelism, which equates with a larger circuit. Therein lies the basic conceptual trade-off between speed and area: Faster circuits require more parallelism and an increase in area. Because of the second-order effects from FPGA layout, however, this does not always work out as expected.

It isn't until place and route is completed before the tool really knows how congested the device is or the difficulty in the place and route process. At this point in the flow, a particular logic topology has already been committed to by the synthesis tool. Thus, if an optimization effort was set to speed at the synthesis level and the back-end tool finds that the device is overly congested, it must still attempt to place and route all the extra logic. When the device is congested, the tool will have no choice but to place the components wherever they will fit and will therefore introduce long delays due to the suboptimal routes. Because of the fact that designers will often use the smallest FPGA possible for economic reasons, this situation occurs very frequently. This leads to the general heuristic:

As the resource utilization approaches 100%, a speed optimization at the synthesis level may not always produce a faster design. In fact, an area optimization can actually result in a faster design.

The plot in Figure 14.1 represents actual timing data from a RISC microprocessor implemented in a Virtex-II FPGA relative to the timing constraint.

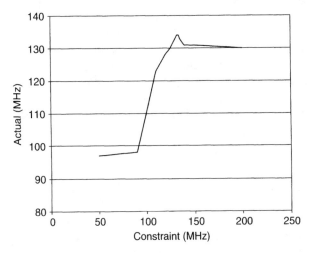

Figure 14.1   Actual speed versus constrained speed.

To illustrate the effect, the critical path was constrained to a tight physical region via placement constraints.

Based on Figure 14.1, the performance graph can be divided into four regions:

- *Underconstrained*: This is the flat region near the bottom where the constraint was defined to be less than 95 MHz. In this region, a compact implementation of the logic will run at approximately 95 MHz without any significant timing optimizations.

- *Optimization region*: The linear region between 95 MHz and 135 MHz that represents the range where increases in the timing constraints can be fulfilled by corresponding optimizations in the logic implementation. In other words, the timing constraints in this region can be met by synthesizing higher speed (and correspondingly higher area) logic structures.

- *Peak*: The upper peak represents the maximum constraint that can be satisfied with improvements to the logic structure given the parallel architectures for the specific design and the amount of space available in the FPGA.

- *Overconstrained*: This is the flat region near the top where the constraint exceeds the maximum achievable frequency.

This example helps to illustrate the problem associated with overconstraining a design. If the target frequency is set too high (i.e., greater than 15–20% above the final speed), the design can be implemented in a suboptimal manner that actually results in a lower maximum speed. During the initial implementation, the synthesis tool will create logic structures based on the timing requirements. If during the initial timing analysis phase it is determined that the design is too far from achieving timing, the tool may give up early. If, however, the constraints are set to the correct target and not more than 20% above the final frequency (assuming it does not make timing initially), the logic will be implemented with minimal area to achieve the specified timing and will have more flexibility during timing closure. It is also important to point out that due to second order effects of the FPGA implementation, targeting a smaller design may or may not also improve timing depending on the particular circumstances. This issue is discussed in later sections.

14.2 RESOURCE SHARING

Architectural resource sharing was discussed in an earlier chapter, whereby portions of the design that can be used for different blocks of functionality are reused via steering logic. At a high level, this type of architecture can dramatically reduce the overall area with a penalty that may include throughput if the operations are not mutually exclusive. Resource sharing on the synthesis optimization level typically operates on groups of logic between register stages. These simpler architectures can be boiled down to simple logical and often arithmetic operations.

A synthesis engine that supports resource sharing will identify similar arithmetic operations that are mutually exclusive and combine the operations via steering logic. For instance, consider the following example.

```
module addshare (
  output oDat,
  input iDat1, iDat2, iDat3,
  input iSel);

  assign oDat = iSel ? iDat1 + iDat2: iDat1 + iDat3;
endmodule
```

In the above example, the output oDat is assigned either the sum of the first two inputs or the sum of the first and third input depending on a selection bit. A direct implementation of this logic would be as shown in Figure 14.2.

In Figure 14.2, both sums are computed independently and selected based on the input iSel. This is a direct mapping from the code but may not be the most efficient method. An experienced designer will recognize that the input iDat1 is used in both addition operations, and that a single adder could be used with the inputs iDat2 and iDat3 muxed at the input as shown below.

This result can also be achieved by use of a synthesis-provided resource sharing option. Resource sharing will identify the two add operations as two mutually exclusive events. Either one adder will be updated or the other depending on the state of the selection bit (or other conditional operator). The synthesis tool is then able to combine the adders and mux the input (Fig. 14.3).

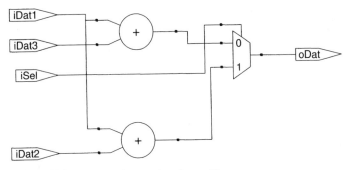

Figure 14.2   Direct implementation of two adders.

Although the maximum delay through the above implementation was not affected by the resource-sharing optimization, there are cases where resource-sharing will require additional muxing along an individual path. Consider the following extension of the previous example.

```
module addshare (
  output      oDat,
  input       iDat1, iDat2, iDat3,
  input [1:0] iSel);

  assign oDat = (iSel == 0) ? iDat1 + iDat2:
                (iSel == 1) ? iDat1 + iDat3:
                iDat2 + iDat3;

endmodule
```

A direct mapping will produce a structure as shown in Figure 14.4. This implementation has been created with parallel structures for all adders and selection logic. The worst-case delay will be the path through an adder plus a mux. With resource sharing enabled, the adder inputs are combined as shown in Figure 14.5. In this implementation, all adders have been reduced to a single adder with muxed inputs. Note now, however, that the critical path spans three levels of logic. Whether or not this actually affects the timing of this path depends not only on the specifics of the logic that is implemented but also on the available resources in the FPGA.

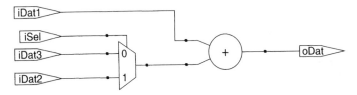

Figure 14.3   Combined adder resource.

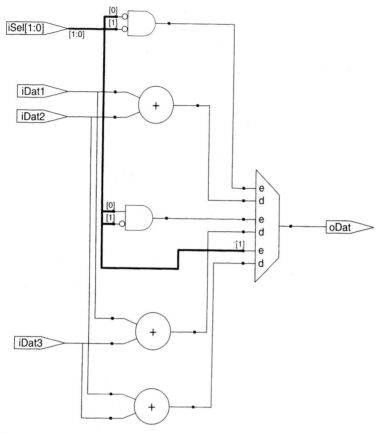

Figure 14.4 Direct mapping of three adders.

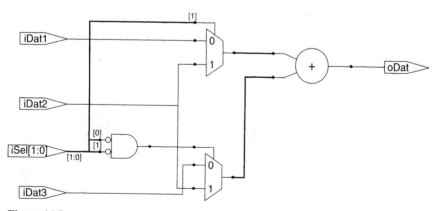

Figure 14.5 An extra logic level when adders are shared.

An intelligent synthesis tool will typically utilize resource sharing if the path is not critical; that is, if the operation is not in the worst-case flip-flop to flip-flop timing path. If the synthesis tool has this capability, then this option will almost always be useful. If not, then the designer must analyze the critical path to see if this optimization is adding additional delay.

If resource sharing is activated, verify that it is not adding delay to the critical path.

14.3 PIPELINING, RETIMING, AND REGISTER BALANCING

In earlier chapters that discussed architecting for speed, pipelining was a method that was used to increase the throughput and flip-flop to flip-flop timing by adding register stages between groups of logic. A well-designed module can usually be pipelined by adding additional register stages and only impact total latency with a small penalty in area. The synthesis options for pipelining, retiming, and register balancing operate on the same structures but do not add or remove the registers themselves. Instead, these optimizations move flip-flops around logic to balance the amount of delay between any two register stages and therefore minimize the worst-case delay. Pipelining, retiming, and register balancing are very similar in meaning and often only vary slightly from vendor to vendor. Conceptually, this is illustrated in Figure 14.6.

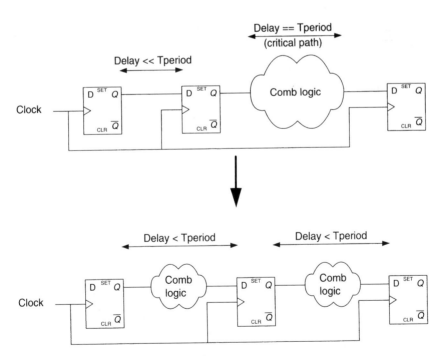

Figure 14.6 Balancing combinatorial logic.

Pipelining typically refers to the first widely adopted method of load balancing whereby regular structures such as pipelined memories or multipliers could be identified by the synthesis tool and rearchitected with redistributed logic. In this case, pipelining requires that a regular pipeline exists and that it is easily recognizable by the tool. For instance, the following code defines a parameterizable pipelined multiplier:

```
module multpipe #(parameter width = 8, parameter depth = 3) (
  output [2*width-1: 0] oProd,
  input  [width-1: 0]   iIn1, iIn2,
  input                 iClk);
  reg    [2*width-1: 0] ProdReg [depth-1: 0];
  integer               i;

  assign oProd      = ProdReg [depth-1];

  always @(posedge iClk) begin
    ProdReg[0]      <= iIn1 * iIn2;

    for(i=1;i <depth;i=i+1)
        ProdReg[i]      <= ProdReg [i-1];
  end
endmodule
```

In the above code, the two inputs are simply multiplied together, registered, and succeeded with a number of register stages defined by the parameter depth. A direct mapping without automatic pipelining will produce the implementation shown in Figure 14.7.

In the example of Figure 14.7, only one register was pulled into the multiplier to serve as an output register (indicated by the number 1 in the multiplier block). The remainder of the pipeline registers are left at the output with an overall imbalance of logic. By enabling the pipeline, we can push the output registers into the multiplier as shown in Figure 14.8. The number "3" in the symbol indicates that the register has a three-layer pipeline internally.

Retiming and register balancing typically refer to the more general case where a flip-flop is moved around logic while maintaining the same logic function to the outside world. This general case is illustrated in the following example.

```
module genpipe (
  output reg  oProd,
```

Figure 14.7 Multiplier behind a pipeline.

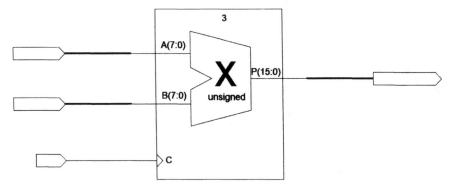

Figure 14.8   Pipeline moved into multiplier.

```
input  [7:0]  iIn1,
input         iReset,
input         iClk);
reg    [7:0]  inreg1;

always @(posedge iClk)
  if(iReset) begin
    inreg1   <= 0;
    oProd    <= 0;
  end
  else begin
    inreg1    <= iIn1;
    oProd <= (inreg1[0]|inreg1[1]) & (inreg1[2]|inreg1[3]) &
             (inreg1[4]|inreg1[5]) & (inreg1[6]|inreg1[7]);
  end
endmodule
```

A synthesis run that is accurate register-for-register would produce the implementation shown in Figure 14.9.

Here, all logic is contained between two distinct register stages as represented in the code. If register balancing is enabled, the overall timing can be improved by moving registers into the critical path logic as shown in Figure 14.10.

As can be seen from the diagram of Figure 14.10, there is no penalty in terms of latency or throughput when register balancing is used. The register utilization may increase or decrease depending on the application, and run time will be extended. Thus, if retiming is not a necessity for timing compliance, an intelligent synthesis tool will not perform these operations on noncritical paths.

Register balancing should not be applied to noncritical paths.

14.3.1 The Effect of Reset on Register Balancing

As with many other optimizations, the reset can have a direct impact on the ability of the synthesis tool to use register balancing. Specifically, if two flip-flops

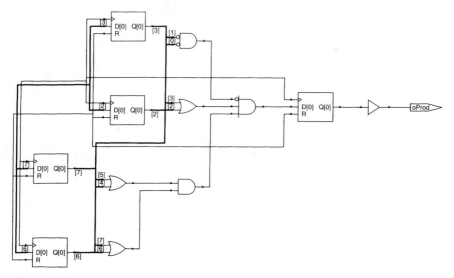

Figure 14.9   Imbalanced logic.

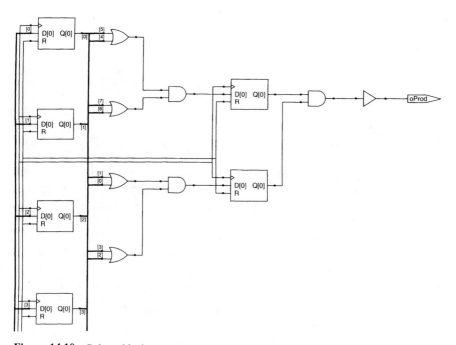

Figure 14.10   Balanced logic.

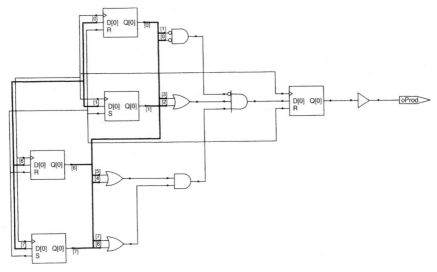

Figure 14.11   Mixed reset types preventing register balancing.

are required to combine to balance the logic load, the two flip-flops must have the same reset state. For instance, if one reset has a synchronous reset and another an asynchronous reset (which would typically be poor design practice) or if one had a set versus a reset, the two could not be combined, and register balancing would have no effect. If this condition existed in the previous example, the final implementation would be prevented as shown in Figure 14.11. In this implementation, the registers that drive the logic gates are initialized to an alternating 1-0-1-0 pattern. This prevents any register balancing or recombination due to incompatible register types. A smart synthesis tool may analyze the path and determine that inverting the reset type along with a corresponding inversion of the flip-flop input and inversion of the flip-flop output (a workaround for the offending flip-flop) will improve the overall timing. However, if the delay introduced by this workaround eliminates the overall effectiveness of the register balancing, the synthesis tool will use a direct mapping and this technique will offer no significant optimization.

Adjacent flip-flops with different reset types may prevent register balancing from taking place.

14.3.2   Resynchronization Registers

One case where register balancing would be a problem is in the area of signal resynchronization. In previous chapters, we discussed the double-flop method for resynchronizing an asynchronous signal either from outside the FPGA or from another clock domain as shown in Figure 14.12.

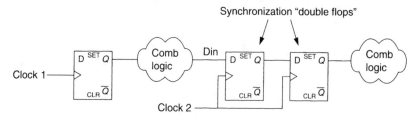

Figure 14.12   Resynchronization registers without balancing.

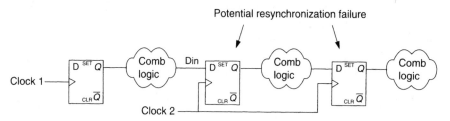

Figure 14.13   Balancing applied to resynchronization registers.

If register balancing is enabled, the logic following the resynchronization registers could be pushed between these registers as shown in Figure 14.13.

Because it is desirable not to perform any logical operations on a potentially metastable signal, as well as provide as much time as possible for the signal to become stable, it is important that register balancing does not affect these special circuits. If register balancing is enabled, these circuits must be analyzed to ensure there will be no effect to the resynchronization. Most synthesis tools will have the ability to constrain the design to prevent register balancing on individual registers.

> Constrain resynchronization registers such that they are not affected by register balancing.

14.4 FSM COMPILATION

FSM compilation refers to the automatic identification of a finite state machine in the RTL and recoding as needed for the speed/area constraints. This means that as long as a standard state-machine architecture is used, the exact coding in the RTL is unimportant. Due to the regular structure of a state machine coded with a standard style, the synthesis tool can easily extract the state transitions and output dependencies and transform the FSM into something that is more optimal for a given design and set of constraints.

Design state machines with standard coding styles so they can be identified and reoptimized by the synthesis tool.

Binary and sequential encoding will depend on all flip-flops in the state representation, and thus a state-decode will be necessary. FPGA technologies that are logic rich or that have multiple input gates for the decode logic will optimally implement these FSMs.

One-hot encoding is implemented such that one unique bit is set for each state. With this encoding, there is no state decode and the FSM will usually run faster. The disadvantage is that one-hot encodings typically require many registers.

Gray codes are a common alternative to one-hot encoding in two primary applications:

- Asynchronous outputs
- Low-power devices

If the output of the state machine, or any of the logic that the state machine operates on, is asynchronous, gray codes are typically preferred. This is due to the fact that asynchronous circuits are not protected from race conditions and glitches. Thus, the path differential between two bits in the state register can cause unexpected behavior and will be very dependent on layout and parasitics. Consider the output encoding for a Moore machine as shown in Figure 14.14. In this case, state transition events will occur where a single bit will be cleared and a single bit will be set, thereby creating the potential for race conditions. The waveforms illustrating this condition are shown in Figure 14.15.

One solution to this problem is to use gray encoding. A gray code only experiences a single bit change for any transition. The fact that gray codes can be used to safely drive asynchronous outputs is apparent after analyzing the structure of coding scheme. To construct a gray code, use the mirror-append sequence as described below.

1. Begin with a "0" and a "1" listed vertically.
2. Mirror the code from the bottom digit.
3. Append "0" to the upper half of the code (the section that was copied in the mirror operation).
4. Append "1" to the lower half of the code (the section that was created in the mirror operation).

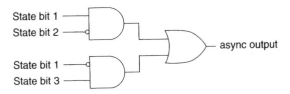

Figure 14.14   Example moore machine output.

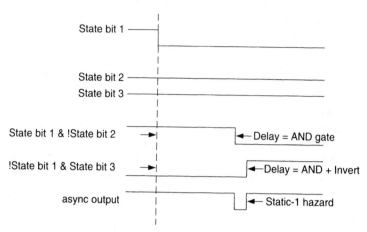

Figure 14.15 Potential hazard.

This sequence is illustrated in Figure 14.16.

As can be seen in Figure 14.16, a gray code will only experience a single bit toggle for every state transition and thus eliminates race conditions within the asynchronous logic.

Use gray codes when driving asynchronous outputs.

Aside from conditions such as described above, one-hot encoding is typically preferred for FPGA design. This is because FPGAs are register rich, there is no decoding logic required, and because it is usually faster. For these reasons, most state machines will be replaced with a one-hot structure, so for good design practice and to reduce overall run time it is advised to design all state machines with one-hot encoding unless there is a compelling reason to do otherwise.

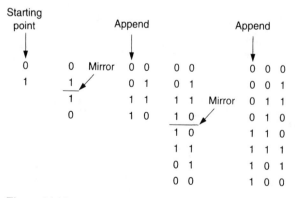

Figure 14.16 Creating gray codes.

14.4.1 Removal of Unreachable States

Most state-machine compilers will remove unused states and may be intelligent enough to detect and remove unreachable states. For most applications, this will help to optimize both speed and reduce area. The main application that would require the retention of unreachable states would be for high-reliability circuits used in aviation, by the military, or in spacecraft. With ultrasmall geometries, particles of radiation from solar or nuclear events can cause flip-flops to spontaneously change states. If this happens in a circuit that is critical to human life, it is important to ensure that any combination of register states has a quick recovery path. If every possible state in a FSM is not accounted for, it is possible that such events could put the circuit into a state from which it cannot recover. Thus, synthesis tools often have a "safe mode" to cover all states even if they are unreachable through normal operation.

The following module contains a simple FSM that continuously sequences through three states after reset. The output of the module is simply the state itself.

```
module safesm (
  output [1:0] oCtrl,
  input        iClk, iReset);
  reg    [1:0] state;

  // used to alias state to oCtrl
  assign oCtrl = state;

  parameter STATE0 = 0,
            STATE1 = 1,
            STATE2 = 2,
            STATE3 = 3;

  always @(posedge iClk)
    if(!iReset) state <= STATE0;
    else
      case(state)
        STATE0: state <= STATE1;
        STATE1: state <= STATE2;
        STATE2: state <= STATE0;
      endcase
endmodule
```

The implementation is simply a shift register as shown in Figure 14.17. Note that if bits 1 and 2 are errantly set simultaneously, this error will continue to recirculate and generate an incorrect output until the next reset. If safe mode is enabled, however, this event will cause an immediate reset as shown in Figure 14.18.

With the implementation of Figure 14.18, an incorrect state will be detected by the additional logic and force the state registers to the reset value.

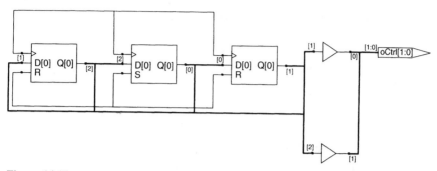

Figure 14.17   Simple state machine implementation.

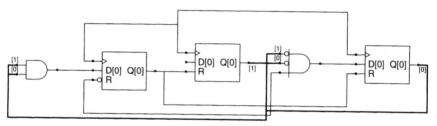

Figure 14.18   State machine implementation with safe mode.

14.5 BLACK BOXES

Black boxes are placeholders in the synthesis flow for a netlist or layout optimized block that will be included in the design later in the implementation flow. The main problem with a black box is that it cannot be optimized by synthesis, and the synthesis tool will have trouble optimizing around it. If the timing engine in the synthesis tool does not know the nature of the interfaces or the timing in and out of the black box, it must assume the worst-case condition and optimize accordingly.

In general, it is not recommended to generate low-level optimized cores that require black boxes in synthesis because this prevents the synthesis from optimizing the entire design. If this is not possible, it is important to include the timing models provided by the core generation tool if optimization around the black box is required.

If a black box is required, include the timing models for the I/O.

Consider the following example of a counter that is periodically pushed into a FIFO.

```
module fifotop(
  output [15:0] oDat,
  output        oEmpty,
```

```
input            iClk, iReset,
input            iRead,
input            iPushReq);
reg    [15:0]    counter;
wire             wr_en, full;

assign wr_en         = iPushReq & !full;

always @(posedge iClk)
  if(!iReset) counter <= 0;
  else        counter <= counter + 1;

myfifo myfifo (
  .clk   (iClk),
  .din   (counter),
  .rd_en (iRead),
  .rst   (iReset),
  .wr_en (wr_en),
  .dout  (oDat),
  .empty (oEmpty),
  .full  (full));
endmodule
```

In the above example, the 16-bit free running counter pushes its current value
into the FIFO whenever a push request is input (iPushReq) and only if the FIFO
is not full. Note that the FIFO myfifo is an instantiation of a core generated from
the Xilinx core generator tool. The black-box instantiation is defined with only
the I/O as placeholders as shown in the following module definition.

```
module myfifo(
  output [15: 0] dout,
  output         empty,
  output         full,
  input          clk,
```

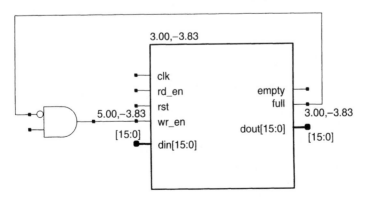

Figure 14.19 Critical path around black box.

```
input  [15: 0]  din,
input           rd_en,
input           rst,
input           wr_en) /* synthesis syn_black_box */;
endmodule
```

The black-box synthesis directive shown in the above code tells the synthesis tool not to optimize this module out of the design and to leave placeholders for the netlist in the back-end tool. The drawbacks as stated earlier are the fact that timing and area utilization are unknown. This prevents the synthesis tool from making any optimizations on or around the interface of the black box. Additionally, if the critical path intersects with the black box, the perceived worst-case timing will be incorrect and will render max speed estimates entirely incorrect. For instance, in the above design, the worst-case path is determined to lie between one of the black-box outputs and one of the inputs.

Because no setup or hold timing is defined for the black box represented in Figure 14.20, this is determined to be zero, and thus a max frequency of 248 MHz is provided as an estimate. To see the inaccuracy of this, consider the modified black-box instantiation that simply defines setup and hold times on the black box relative to the system clock.

```
module myfifo(
    output [15: 0]  dout,
    output          empty,
    output          full,
    input           clk,
    input  [15: 0]  din,
    input           rd_en,
    input           rst,
```

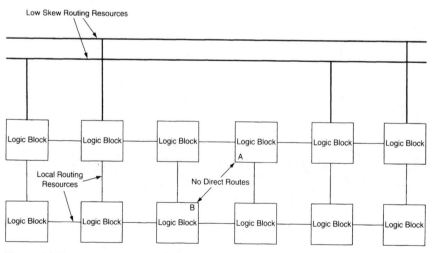

Figure 14.20   Simplified routing matrix.

```
input          wr_en)  /* synthesis syn_black_box
syn_tsu1  =    "din[15:0]->clk=4.0"
syn_tsu2  =    "rd_en->clk=3.0"
syn_tsu3  =    "wr_en->clk=3.0"
syn_tco1  =    "clk->dout=4.0"
syn_tco2  =    "clk->empty=3.0"
syn_tco3  =    "clk->full=3.0"
*/;
endmodule
```

In this case, we define the setup and hold times for the control signals to be 3 ns and the setup and hold times for the I/O data to be 4 ns. Using this new information, the synthesis tool decides that the identical path is still critical, but the timing can now be accurately estimated at 125 MHz, which is about half of the original estimate.

14.6 PHYSICAL SYNTHESIS

Physical synthesis has taken on a number of different forms from various vendors, but ultimately it is the utilization of physical layout information to optimize the synthesis process. In other words, by providing either preliminary physical estimates about placement or by performing actual physical layout, the synthesis tool will have better loading models to work with and will be able to optimize a design with information that is very close to the final implementation. The physical synthesis tools that perform preliminary placement will forward annotate the layout information to the back-end place and route tool.

In some ways, physical synthesis is simpler in the ASIC world than with FPGAs. The reason is than because ASICs have flexible routing resources, the delays can be statistically estimated based on the distance between the two end points of the route. In an FPGA, the routing resources are fixed. There will be a number of fast routing resources (usually long lines between major areas of the chip) and a number of slower routing resources (switch matrices) to handle local routing congestion. Figure 14.20 illustrates a simplified view of an abstract routing matrix.

As can be seen from Figure 14.20, there will be a hierarchy of routing resources. The low-skew lines will have the ability to carry signals long distances around the FPGA with little delay but are very limited in extent among all logic elements. The local routing resources are used to make connections to local resources but are very inefficient in their ability to carry signals for any distance and will be very dependent on congestion in adjacent elements. Thus, the ability to route between point A and point B will greatly depend on the availability of routing resources near the logic elements and the utilization of these resources.

For the reasons discussed above, estimating routing delay can be very complex for FPGA synthesis tools. This is a major problem with modern FPGAs as routing delays become dominant in smaller geometries. Thus, an inaccurate estimate of routing delays can prevent a synthesis tool from optimizing accordingly. Physical synthesis provides a solution to this by generating placement information during synthesis.

14.6.1 Forward Annotation Versus Back-Annotation

Back-annotation is the traditional path to timing closure. The synthesis implements logic based on statistical estimates of the final place and routing timing information. The place and route tool then takes this netlist and creates a physical layout of the logic structures. Timing analysis is then performed, and any timing violations are fed back either to the placement tool (smaller violations can typically be fixed with a better placement as discussed in later chapters) or to the synthesis tool for major violations. In the latter case, the designer creates constraints either by hand to inform the synthesis tool that a particular path needs to be tightened or in an automated manner with more sophisticated tools.

Although the methodology described above does typically bring the design to timing closure, the primary drawback is the total design time and lack of automation to close the entire loop. The above methodology forces the designer to run the tool many times and for each iteration to feed timing information back to synthesis. Figure 14.21 illustrates the traditional timing closure flow, which includes design constraints, synthesis, place and route, and static timing analysis. Note that the timing information for the design is fed back to an earlier point in the flow for optimization.

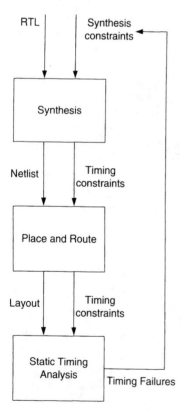

Figure 14.21 Back-annotating timing violations.

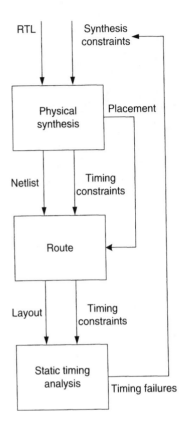

Figure 14.22 Forward-annotating placement data.

Forward annotation, on the other hand, passes design-related information forward as a set of constraints or physical parameters that indicates or defines the assumptions that were made by synthesis. By using this information to drive the back-end implementation, the estimates from synthesis will be more accurate, and the need to re-run the flow will be minimized.

Physical synthesis provides tighter correlation between synthesis and layout.

In Figure 14.22, the timing failures may still need to feed back to the physical synthesis tool. Even if timing failed due to poor routing, either the placement or the fundamental logic structures will need to be reoptimized. The key here is that because synthesis and placement are tightly coupled, the number of iterations before achieving timing closure will be far fewer. The degree to which this process can be automated is directly related to algorithms used in the physical synthesis for placement estimates as discussed in the next section.

14.6.2 Graph-Based Physical Synthesis

One of the keys to high-performance physical synthesis is to create accurate estimates of routing congestion and delay. Synplicity has developed a technology for FPGA physical synthesis called graph-based physical synthesis. The idea is to

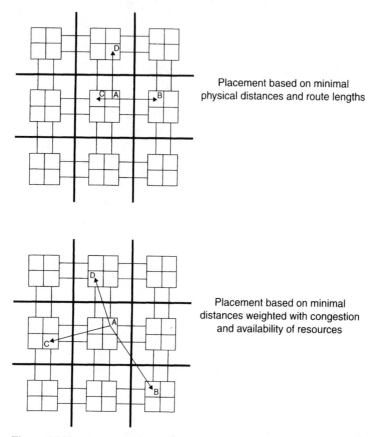

Placement based on minimal
physical distances and route lengths

Placement based on minimal
distances weighted with congestion
and availability of resources

Figure 14.23 Graph based physical synthesis.

abstract the routing resources and related timing information into a routing resource graph. The physical placement is performed based not on a physical distance but on weighted calculations from the graph. Figure 14.23 illustrates the difference between minimum physical distance and weighted distance based on congestion.

The graph-based synthesis flow takes into account certain criteria that will affect the final layout including routing resources between any two points as well as existing utilization. This methodology creates a tighter correlation between synthesis and layout but does not address high-level partitioning issues or critical path constraints. This is discussed further in Chapter 15.

14.7 SUMMARY OF KEY POINTS

- As the resource utilization approaches 100%, a speed optimization at the synthesis level may not always produce a faster design. In fact, an area optimization can actually result in a faster design.

- If resource sharing is activated, verify that it is not adding delay to the critical path.
- Register balancing should not be applied to noncritical paths.
- Adjacent flip-flops with different reset types may prevent register balancing from taking place.
- Constrain resynchronization registers such that they are not affected by register balancing.
- Design state machines with standard coding styles so they can be identified and reoptimized by the synthesis tool.
- Use gray codes when driving asynchronous outputs.
- If a black box is required, include the timing models for the I/O.
- Physical synthesis provides tighter correlation between synthesis and layout.

Chapter 15

Floorplanning

As discussed in Chapter 14, there are only so many optimizations that can be made in the synthesis tool, and only a small subset of those can be forward annotated to the back-end place and route tool such as described in the physical synthesis flow. All of those optimizations are performed at a low level of abstraction and operate on individual logic structures. There are no methodologies discussed thus far that address higher-level constraints that can be passed to the back-end tools to optimize the speed and quality of the place and route algorithms. This chapter describes one such high-level method called floorplanning.

During the course of this chapter, we will discuss the following topics:

- Partitioning a design with a floorplan.
- Performance improvements by constraining the critical path.
- Floorplanning dangers.
- Creating an optimal floorplan.

   Floorplanning the data path
   Constraining high fan-out logic
   Shaping the floorplan around built-in FPGA devices
   Reusability

- Floorplanning to reduce power dissipation.

15.1 DESIGN PARTITIONING

As device densities and corresponding design sizes have become very large (millions of gates), newer methodologies have been developed to assist the placement tools when laying out logic elements from a sea of gates. To address this issue, floorplanning has become common in the area of design partitioning. A typical design flow that utilizes floorplanning to partition a design is shown in Figure 15.1 (with and without physical synthesis).

Advanced FPGA Design. By Steve Kilts
Copyright © 2007 John Wiley & Sons, Inc.

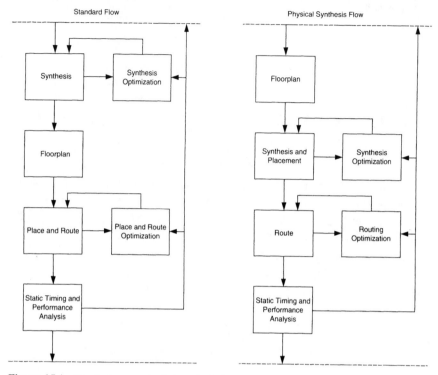

Figure 15.1 Standard versus physical synthesis flow.

In the two flow diagrams of Figure 15.1, the floorplanning stage must always come before the placement operation. In the case of the physical synthesis flow, the floorplan is created prior to synthesis. In either case, the floorplan is fixed and used as a guide to partition the physical location of all logic elements.

A high-level floorplan such as one that is used to partition designs not only captures all major blocks in the design but also abstracts those blocks such that the individual logic structures need no specific reference. This allows large functional blocks to be partitioned for very large designs that have precisely defined timing at their interfaces. For instance, consider a design with three main functional blocks called A, B, and C, respectively. The pipeline is defined as shown in Figure 15.2.

In the design partition of Figure 15.2, there are three major functional blocks, each of which is assigned to a separate designer. The blocks only interact at specific interfaces that have been predefined by the system designer, who has also defined the timing of these interfaces by means of a timing budget. Because the interfaces have been clearly defined, and most likely all I/O of the functional blocks will be registered, the critical path timing will lie within the blocks themselves. In other words, with a good design partition, the critical path will not cross any major functional boundaries, and so timing compliance can be

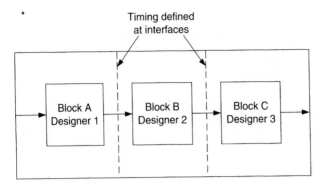

Figure 15.2 Pipeline with well-defined interfaces.

considered on a block-by-block basis. This assumption can dramatically speed up the implementation of a large device as an alternative to placing a sea of gates.

By partitioning the floorplan between major functional boundaries, timing compliance can be considered on a block-by-block basis.

The floorplanner from Synplicity provides outstanding control over a partitioned design on a very abstract level. Figure 15.3 shows a possible floorplan of the design partition of Figure 15.2 for a Spartan-3 5000.

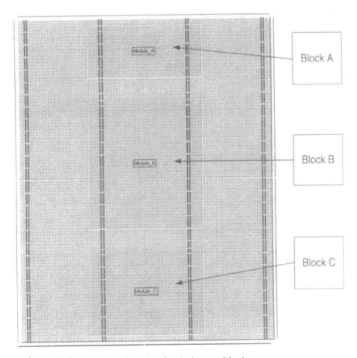

Figure 15.3 Example floorplan for design partitioning.

The regions in this diagram define the physical area occupied by each block. An FPGA device of this size would normally take many hours to successfully place and route with a blanket sea of gates. With the above partitioning, however, this run time will be dramatically reduced to three smaller (and more manageable) place and route operations. Assuming all interfaces are registered at the boundary, the relatively large gaps between the blocks will not cause timing problems. Besides run time, another benefit to this type of partitioning is that major structural or layout changes in one block need not affect the others. Thus, a methodology that employs design partitioning works in close harmony with an incremental design flow.

15.2 CRITICAL-PATH FLOORPLANNING

Floorplanning is often used by designers who have very difficult timing constraints and need to tighten their critical paths as much as possible. The floorplan in this case would be created after the final implementation results were generated and the critical path was defined. This information would be back-annotated to the floorplanner whereby the designer would manually define location constraints for the critical logic elements. These physical constraints would then be forward annotated into the place and route tool to complete an iterative cycle. Figure 15.4 illustrates the design flow when using a floorplan to constrain only critical paths.

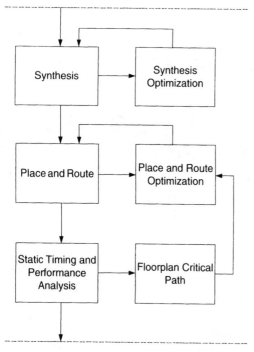

Figure 15.4   Design flow with critical path floorplanning.

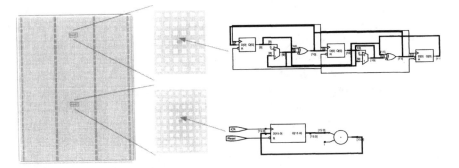

Figure 15.5   Example critical-path-floorplan constraints.

In this case, the floorplanning step is not a static and unchanged step in the process as it is with a design partition but rather a key link in the iterative timing closure loop.

When floorplanning the critical path, the floorplan is a key link in the iterative timing closure loop.

Note that the floorplan is not modified until a critical path has been determined and will be modified during every iteration of the timing closure loop. Figure 15.5 illustrates two possible critical-path-floorplan constraints.

In this example, there are no major functional partitions in the floorplan. Instead, the floorplan consists of two small regions designed to tighten the timing for two distinct critical paths in the design. These regions are created *ad hoc* based on the timing info that is fed back to the floorplanner and will be updated for every iteration of the timing closure loop.

15.3   FLOORPLANNING DANGERS

The danger in floorplanning is that if done incorrectly, it can dramatically *decrease* the performance of the implementation. This is because of the fact that good placement directly corresponds with a design that performs well, and bad placement directly corresponds with a design that performs poorly. This may seem like an obvious statement, but a corollary to this is that a bad floorplan will lead to bad placement and will subsequently lead to poor performance. Thus, a floorplan of any type will not have a nondecreasing impact on performance. Rather, a bad floorplan will make things much worse.

A bad floorplan can dramatically reduce the performance of a design.

It is important to note that not all designs lend themselves to floorplanning. Designs that are pipelined and have a very regular data flow, such as with a pipelined microprocessor, clearly lend themselves to floorplanning. Devices that implement primarily control or glue logic or that don't have any definable major data path often do not lend themselves to a good floorplan that is intended to partition the design. If the design is indeed simply a sea of gates, then it would be optimal to allow the synthesis and place and route tools to treat it as such.

One general heuristic to determine if a design is a good candidate for critical-path floorplanning is to analyze the routing versus logic delays. If the percentage of a critical path that is consumed in routing delay is the vast majority of the total path delay, then floorplanning may assist in bringing these structures closer together and optimizing the overall routing resources and improving timing performance. If, however, the routing delay does not take up the majority of the critical-path delay and there is no clearly definable data path, then the design may not be a good candidate for floorplanning.

> Floorplanning is a good fit for highly pipelined designs or for layouts dominated by routing delay.

For designs that may lend themselves to a good floorplan, there are a number of considerations that must be taken into account to ensure that the performance will actually be improved. This is discussed in the next section.

15.4 OPTIMAL FLOORPLANNING

The optimal floorplan will group logic structures that have direct interconnect in close proximity to one another and will not artificially separate elements that may lie in a critical path. The following sections describe methods to optimize the floorplan.

15.4.1 Data Path

A design that is data path centric is often relatively easy to partition. For most high-speed applications, the pipeline to be partitioned will usually apply to the data path. Because the data path carries the processed information and is required to do so at very high speeds (often running continuously), it is recommended to floorplan this first as shown in Figure 15.6.

In this scenario, a floorplan is created to partition the data path only. This includes the main pipeline stages and all associated logic. The control structures and any glue logic that do not lie on the primary data path can be placed automatically by the back-end tools.

> The floorplan usually includes the data path but not the associated control or glue logic.

15.4.2 High Fan-Out

High fan-out nets are often good candidates for floorplanning as they require a large amount of routing resources in one specific region. This requirement often

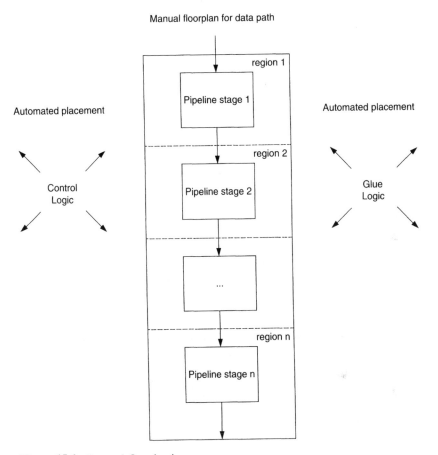

Figure 15.6   Data-path floorplanning.

leads to a high degree of congestion if the loads are not placed in a close proximity to the driver as shown in Figure 15.7.

If the loads are located a relatively long distance from the driver, the interconnect will occupy a large amount of routing resource at the output of the driver. This will make other local routes more difficult and correspondingly longer with larger delays. Figure 15.8 illustrates the benefit of constraining the high fan-out region to a small area.

By confining the high fan-out region to a small and localized area, the effects to other routes will be minimized. This will provide faster run times for the place and route tool as well as a higher performance implementation due to the minimization of routing delays.

15.4.3 Device Structure

The device structure is critical in the floorplan, as built-in structures cannot be moved around with either the floorplan or the placement tools. These built-in

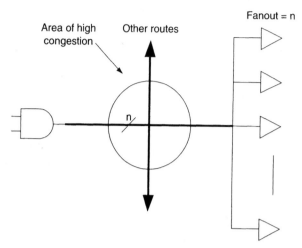

Figure 15.7   Congestion near high fan-out regions.

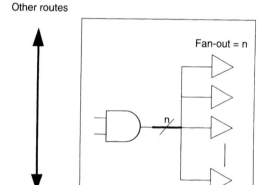

Figure 15.8   Floorplanning high fan-out regions.

structures include memories, DSPs, hard PCI interfaces, hard microprocessors, carry chains, and so forth. It is therefore important not only to floorplan the design such that the blocks are placed optimally relative to one another but also such that built-in structures can be utilized efficiently and that routes to the custom logic are minimized.

> A floorplan should take into consideration built-in resources such as memories, carry chains, DSPs, and so forth.

In Figure 15.9, input and output logic resources are tied to the RAM interface. For a sizable RAM block, it is usually more desirable to use the fixed RAM

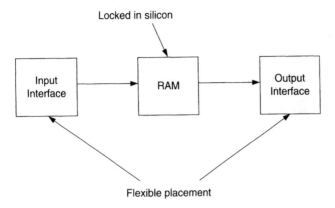

Figure 15.9 Fixed FPGA resources.

resources available in certain regions of the FPGA as opposed to an aggregation of smaller RAM elements distributed throughout the FPGA. The constraint for a floorplan on this path would then be dependent on the fixed location of the RAM resource as shown in Figure 15.10.

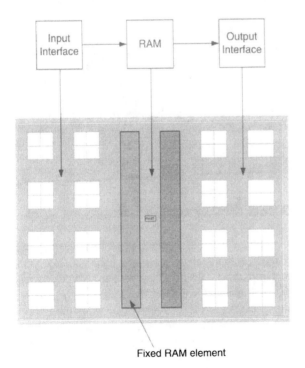

Figure 15.10 Floorplanning around fixed FPGA resources.

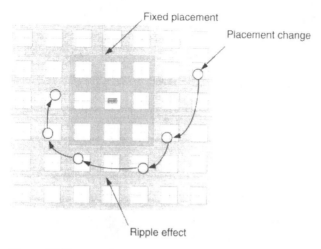

Figure 15.11   Shielding from layout changes.

15.4.4 Reusability

A good floorplan will allow various modules and groups of modules to be reused without a dramatic impact on the performance of those modules. In a sea of gates design, it is common to find that changes to totally unrelated areas of the design can cause timing problems in other aspects of the design. This is due to the progressive nature of an unconstrained place and route algorithm. The chaos effect ensues when a small placement change on one side of the FPGA shifts logic resources over, forcing other local logic structures out of the way, and so on until the entire chip has been replaced with entirely new timing. For a constrained placement via a good floorplan, this is not an issue as the relative timing is fixed for the critical-path modules, and any changes must operate around that floorplan.

In Figure 15.11, the critical logic is constrained inside the floorplan region and will not be affected internally as placement changes around it.

15.5 REDUCING POWER DISSIPATION

Previous sections discussed floorplanning in the context of timing performance, organization, and implementation run-time. There is an additional use for floorplanning, and that is the reduction of the dynamic power dissipation. A route from logic device A to device B is illustrated in Figure 15.12.

Because of the fact that the capacitance of the trace (C_{trace}) is proportional to the area of the trace, and assuming the width of the trace is fixed for an FPGA routing resource, the capacitance will be proportional to the length of the trace. In other words, the capacitance that the driver must charge and discharge will be proportional to the distance between the driver and the receiver.

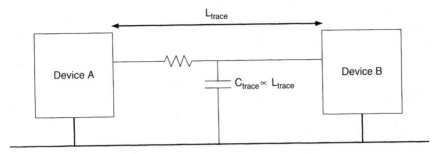

Figure 15.12   LC trace parasitics.

In our original discussion on dynamic power dissipation, it was determined that the power dissipated on high-activity lines was proportional to the capacitance of the trace times the frequency. In this section, we will assume that the functionality is locked down and the activity on the various traces cannot be changed. Thus, to minimize the power dissipation, we must minimize the routing lengths of high-activity lines.

For timing optimization, the place and route tool will place critical-path components in close proximity to one another in an attempt to achieve timing compliance. The critical path, however, does not indicate anything about high activity. In fact, a critical path may have a very low toggle rate, whereas a component that easily meets timing such as a free running counter may have a very high toggle rate. Consider the scenario where the critical path has been constrained to a small region.

In Figure 15.13, the critical path has been constrained, and the clock frequency has been maximized. However, the timing-driven placement has not identified the high-activity circuit as a timing problem and has scattered the components around

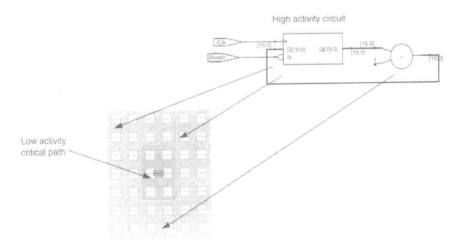

Figure 15.13   Floorplanning to minimizing high-activity nets.

the critical path as was convenient. The problem is that due to the long routes on the high-activity lines, the total dynamic power dissipation will be relatively high. To reduce this power consumption, another placement region can be added for the high-activity circuit to ensure that the interconnect is minimized. If the high-activity circuit is independent of the critical path, this can usually be done with little impact to the timing performance.

A floorplan targeted at minimizing trace lengths of high-activity nets can reduce dynamic power dissipation.

15.6 SUMMARY OF KEY POINTS

- By partitioning the floorplan between major functional boundaries, timing compliance can be considered on a block-by-block basis.
- When floorplanning the critical path, the floorplan is a key link in the iterative timing closure loop.
- A bad floorplan can dramatically reduce the performance of a design.
- Floorplanning is a good fit for highly pipelined designs or for layouts dominated by routing delay.
- The floorplan usually includes the data path but not the associated control or glue logic.
- A floorplan should take into consideration built-in resources such as memories, carry chains, DSPs, and so forth.
- A floorplan targeted at minimizing trace lengths of high-activity nets can reduce dynamic power dissipation.

Chapter 16

Place and Route Optimization

$\mathbf{M}$ost implementation tools for FPGA layout (commonly referred to by the two main stages of layout: place and route) provide the designer with dozens of optimization options. The problem designers will run into with these options is similar to the problem they will run into with synthesis optimization; that is, they typically do not fully understand the implication of all of these options. Therefore, certain optimizations are often used where they are not needed, and other optimizations go unused where they could have a significant impact on performance.

Similar to Chapter 14, this chapter describes the most important aspects of place and route optimization and provides practical heuristics that can be immediately leveraged by the reader. During the course of this chapter, we will discuss the following topics:

- Creating optimal constraints prior to place and route optimization.
- The relationship between placement and routing.
- Reducing route delays with logic replication.
- Optimization across hierarchy.
- Packing registers into I/O buffers.
- Utilizing the pack factor.
- When to pack logic into RAM.
- Ordering registers.
- Variances with placement seed.
- Greater consistency with guided place and route.

16.1 OPTIMAL CONSTRAINTS

Before launching into the actual place and route optimizations themselves, it is important to emphasize the importance of defining a complete set of timing

Advanced FPGA Design. By Steve Kilts
Copyright © 2007 John Wiley & Sons, Inc.

constraints. The reason this topic is listed first is to ensure that the optimizations contained in this chapter are not addressed until sufficient time has been spent creating accurate design constraints.

A complete set of constraints should be created before any optimization takes place.

The constraints that must be included on every design include all clock definitions, I/O delays, pin placements, and any relaxed constraints including multicycle and false paths. Although a particular path may not be a critical timing path, often by relaxing specifications on unrelated paths, placement and routing resources can be freed for more critical paths. For more information regarding these constraints, see Chapter 18.

One class of constraints that is often not covered by typical timing analysis for FPGAs (and is not included in Chapter 18) includes voltage and temperature specifications. The voltage/temperature specifications are among the set of constraints that most often go overlooked by designers but in many cases provide the easiest way to achieve significant timing improvements.

Most designers know that all FPGA devices (like most semiconductor devices) specify a worst-case voltage and temperature under which the device will operate. In the context of FPGA timing analysis, worst-case temperature would be the highest temperature, and the worst-case voltage would be the lowest voltage, because both of these constraints will increase the propagation delays (we typically don't worry about hold delays in FPGAs because of the minimum delays built into the routing matrix). For example, the high end of the commercial temperature range for Xilinx FPGAs is 85°C, and the worst-case voltage rating is usually between 5% and 10% of the recommended operating voltage. This would correspond with approximately 1.14 V for a 1.2-V rail, 3.0 V for a 3.3-V rail, and so on.

These worst-case voltage and temperature conditions are used by default when performing worst-case timing analysis. However, few systems will require the FPGA to run at the extreme conditions of 85°C (junction temperature of the FPGA) and a 10% droop in the power rail. In fact, most systems will only experience conditions far less extreme than this.

If the system in which the FPGA operates is designed within certain temperature and voltage conditions, then the FPGA timing analysis can and should be performed under the same conditions.

When this is taken into consideration, many engineers find that they can shave full nanoseconds off of their critical path (depending on the device technology of course). If the timing violations are less than this incremental improvement, engineers will find that they were struggling over timing violations that didn't really exist!

Consider the example of a 16-bit counter. Under default timing analysis conditions, this counter will run at 276 MHz in a Virtex-II device. The menu in

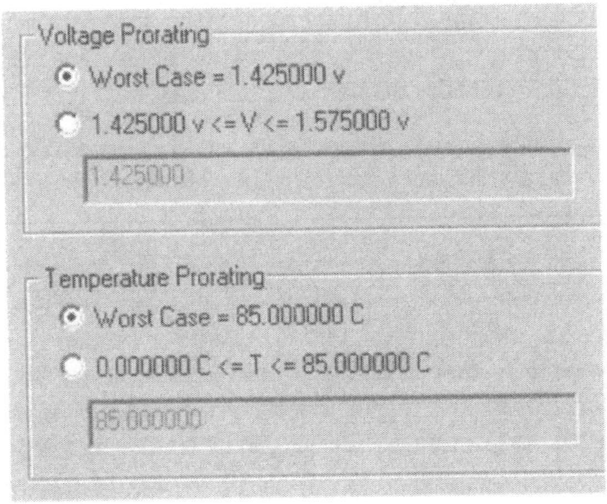

Figure 16.1   Default operating voltage and temperature settings.

Figure 16.1 shows the default voltage and temperature settings that will be passed to the Xilinx timing analysis tool.

In this case, the worst-case voltage setting is 95% of the specified supply voltage. A 5% voltage variance may be reasonable for a switching supply, but if this is running off of a linear supply that can supply more than enough current to the core, this specification is overkill. Additionally, the junction temperature setting is at 85°C. Most likely, the functional design will not generate such a temperature, and thus this specification is again probably overkill. We can reduce these to more reasonable parameters for our hypothetical "light-weight" system where the supply only varies by 2% and the temperature never exceeds 45°C (Fig. 16.2).

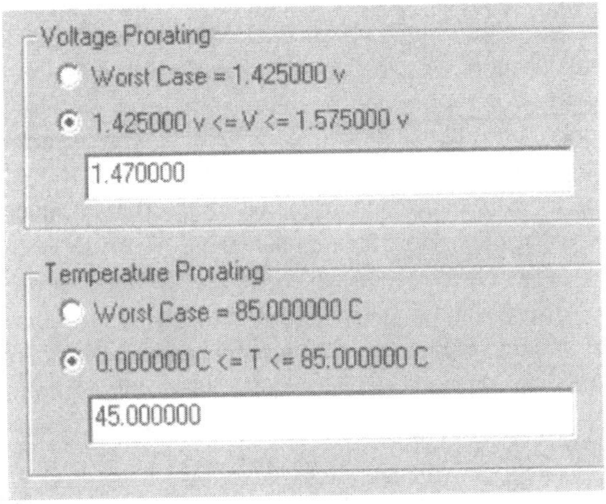

Figure 16.2   Realistic operating voltage and temperature settings.

The result is that the counter now runs at just under 290 MHz. This performance improvement did not require any actual changes to the FPGA but was simply a matter of analyzing the design under the correct conditions.

Adjusting the voltage and temperature settings does not require any changes to the FPGA implementation and can provide an easy means to incrementally improve the worst-case performance.

The only drawback to this approach is that devices need to be fully characterized according to temperature and voltage before the vendor will allow access to this option. Due to the ongoing battle between the FPGA vendors to produce the fastest FPGAs before their competitors, they typically release devices before this characterization has been completed. In fact, if you are using a brand-new technology, you can usually count on this option not being available. In the timing analysis tools, this option will be inaccessible. This is one of the advantages of using a device with a moderate amount of maturity.

16.2 RELATIONSHIP BETWEEN PLACEMENT AND ROUTING

Most modern FPGA place and route tools do a very good job of achieving timing compliance with the automated algorithms, and because of this, manual placement and/or routing is no longer commonplace. However, many FPGA design engineers have been "spoiled" by automated implementation tools to the point where the layout is not even considered outside of very-high-speed designs. This lack of attention has created a lack of understanding and therefore a very inefficient use of the various implementation options available for the layout. Most important of all these options, whether the goal is area or speed, is the relationship of placement and routing relative to processor effort.

Specifically, we are referring to the hooks provided to the FPGA design in the back-end tools that allow the designer to adjust the processor effort and corresponding algorithmic sophistication of the placement and routing. These options are almost always presented and controlled independently, which is unfortunate due to the strong relationship they possess.

In basic training, engineers are told that they can increase the effort of placement and routing to achieve better results. We practice increasing the effort levels in lab experiments and see the improvements. In practice, we see similar improvements. So what's the problem? The problem is that if you turn the router effort level up beyond trivial effort before you have an optimal placement, you are wasting time. Because of the fact that most projects have schedules with deadlines and the fact that most FPGA designers go through many iterations before achieving timing closure, this is certainly something worth considering.

The fundamental concept to understand here is that routing is extremely dependent on placement. For a design of any complexity, a good route can only be achieved with a good placement. Most FPGA design engineers have not run enough experiments to realize this on their own, but placement is vastly more important to performance than is routing. This is particularly true of FPGAs due to the coarse nature of a routing matrix (in an ASIC, there is more flexibility to be creative with the routing).

If you were to take a typical design and run hundreds of combinations of placement and routing effort levels and then plot your data, you would likely see something very similar to the graph shown in Figure 16.3.

As can be seen from the curves in Figure 16.3, placement has a dominant (call it first-order effect) on the performance of the design, and routing has a relatively minor effect.

Placement effort has a dominant effect on performance, whereas routing effort has a relatively minor effect.

To determine the optimal placement and routing effort levels, a designer should follow this procedure:

1. Set both placement and routing effort levels to minimum.

2. Run place and route to determine if the worst case timing is in fact due to a sub-optimal layout and not due to excessive logic levels.

3. Increment placement effort until timing is met or max effort has been reached.

4. If at max placement effort timing is not met, begin incrementally increasing the routing effort.

5. If timing closure cannot be met, revisit the architecture of the design.

6. If a high routing effort is required to meet timing, it may be a good idea to revisit the architecture of the design to optimize for timing (see Chapter 1).

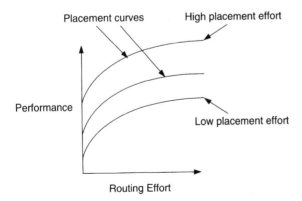

Figure 16.3   Performance versus placement and routing effort.

16.3 LOGIC REPLICATION

Logic replication occurs early in the placement process for structures that fan out to other logic elements that cannot (for any number of reasons) reside in the same proximity. The problem this addresses is illustrated in Figure 16.4.

In this scenario, the output of D2 fans out to two structures that are relatively distant from one another. Regardless of where the driver is placed, the resultant route will be lengthy to one of the driven elements. To eliminate potentially long routing delays, logic duplication will replicate the driver as shown in Figure 16.5.

> Logic duplication should be used only on critical path nets with multiple loads that cannot be physically localized.

The effect of this duplication is that an individual driver can be placed closer to each load, thereby minimizing the route length and delay of the route. Clearly,

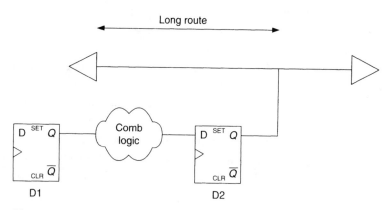

Figure 16.4   A fan-out that forces long routes.

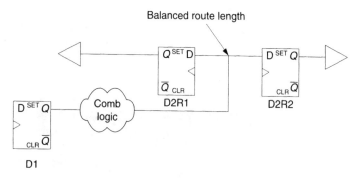

Figure 16.5   Registers duplicated to balance route length.

this optimization will reduce routing delays but will increase area. This optimization may cause inferior results if the utilization of the device is already high. If the placement tool is not intelligent enough to only perform this on critical-path routes, this option may need to be disabled and duplication added to the RTL with corresponding "don't touch" attributes to ensure that the synthesis does not optimize the duplicate structures.

16.4 OPTIMIZATION ACROSS HIERARCHY

Optimizing across hierarchical boundaries will allow any of the placement algorithms to operate when the path extends across a module boundary as shown in Figure 16.6.

Typically, if a particular optimization is desired, it will be beneficial to apply it to intermodule interfaces as well as simply the paths within the modules themselves. Often, logic at the module boundaries does not fully occupy entire LUTs. An example is shown in Figure 16.7.

In this above, a separate LUT is used to implement a NAND operation at each module boundary. By enabling the optimization across boundaries, each of these logical operations can be combined into a single LUT, thereby reducing the area utilization of this operation by one half.

The primary case where this would not be desired is if gate-level simulations need to be run on the postimplementation netlist. In this case, a hierarchy that is intact will be very desirable.

Optimization across hierarchy is not desirable when gate-level simulations are required.

If debug is required on the back-annotated netlist, a preserved hierarchy will allow the designer not only to traverse the design but also to easily identify signals at the boundaries of the modules that may be useful for debug.

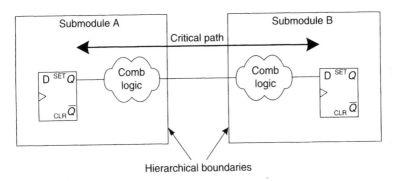

Figure 16.6   Critical path that crosses hierarchy.

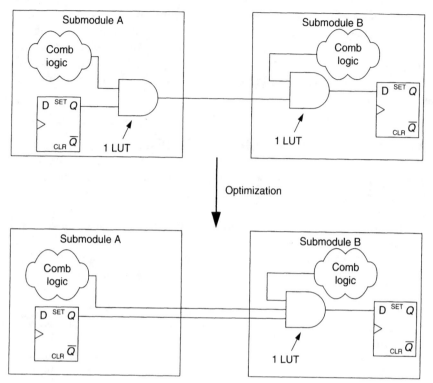

Figure 16.7 Example hierarchy optimization.

16.5 I/O REGISTERS

Many FPGAs have flip-flops built into the input and output buffers to optimize the timing in and out of the chip. Along with these special I/O buffers will be an optimization to enable or disable the packing of these registers into the I/O. Figure 16.8 illustrates the concept of packing registers into the I/O buffers.

There are a number of advantages to placing a register in the I/O:

- The delays at the I/O of the FPGA are minimized.
- More logic is available internally.
- Superior clock-to-out timing.
- Superior setup timing.

The disadvantage of this optimization is that a register that is placed in an I/O buffer may not be optimally placed for the internal logic as shown in Figure 16.9.

For high-speed designs that have tight timing requirements at both the I/O and the internal logic, it may be advantageous to add another layer of pipeline registers at the I/O if allowed by the design protocol as shown in Figure 16.10.

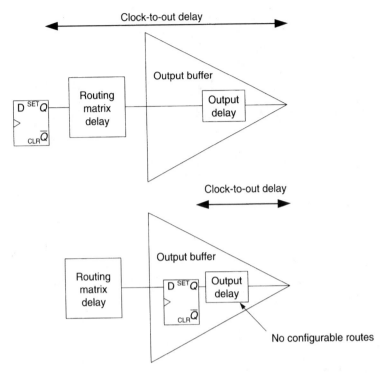

Figure 16.8   Register packed into I/O buffer.

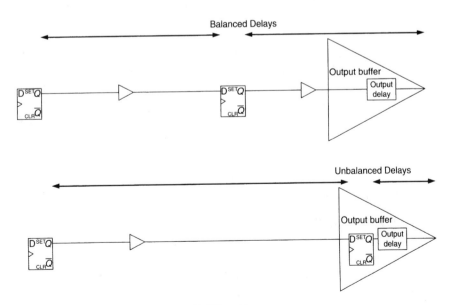

Figure 16.9   Imbalanced route delays with I/O packing.

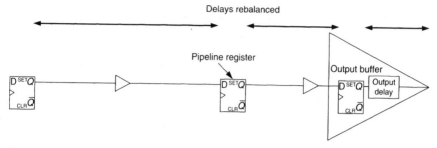

Figure 16.10   Additional pipeline register with I/O packing.

If there are a large number of I/O registers, the extra pipeline layer may add significant overhead in terms of register utilization and potentially congestion.

An extra pipeline register may be required for high-speed designs when packing registers into I/O.

Thus, if there are not tight I/O timing requirements and there are a relatively large number of I/O registers, this optimization is not recommended.

16.6 PACK FACTOR

The pack factor is defined as a percentage and will artificially limit the usage of logic resources in the FPGA. For instance, a pack factor of 100% tells the placement tool that all logic elements are available for implementation, whereas a pack factor of 50% will tell the tool that only half of the total logic resources are available.

The uses for this optimization are limited but can be leveraged by advanced users. For instance, to keep a placeholder for future logic that is not included in the current flow, the pack factor can be reduced according to the estimated size. The ease of implementation will be similar to what the designer can expect when the new core is integrated. Thus, resource utilization problems may be identified sooner.

Additionally, the pack factor can be used to determine the amount of headroom in a design, or "true" utilization. If there are unused logic elements, the place and route tool will be more liberal in the duplication of logic elements and the manner in which these are spread out to optimize the layout. A logic element is defined as utilized if any portion of it is used to implement logic and not necessarily when it is fully utilized. Thus, the percentage utilization will usually be higher than the true utilization.

Setting a pack factor can help to determine the true utilization.

In other words, an FPGA that reports 60% utilization may have much more than 40% more logic resources available (ignoring the routability issues as utilization approaches 100% of course). To estimate true utilization and headroom in the design, the pack factor can be reduced until the design cannot be properly routed.

16.7 MAPPING LOGIC INTO RAM

The major players in the high-end FPGA world are SRAM based, which means that logic functions are coded into LUTs. These LUTs are small SRAM cells distributed across the FPGA and available for general logic implementation. It would seem that a natural extension of this would be to implement logic in the larger dedicated RAM blocks (the ones actually used like RAM), particularly when utilization is running low. Although this may make sense conceptually, the problem associated with this is performance.

The small distributed RAM cells will have very small delays, and logic will propagate through these LUTs very quickly and efficiently relative to other logic elements. The larger RAM blocks, on the other hand, will have much larger delays associated with them and will subsequently create a very slow implementation. In general, it is not wise to rely on logic that is packed into dedicated RAM blocks. Only as a last resort in an extremely high density and slow design would this be useful.

16.8 REGISTER ORDERING

Register ordering is a method used by the placement tool to group adjacent bits of a multibit register into a single logic element. Most cell-based logic elements have more than one flip-flop, and thus by placing adjacent bits together the timing can be optimized as shown in Figure 16.11.

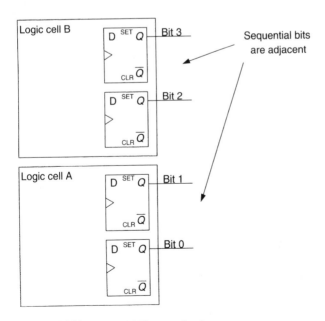

Figure 16.11   Sequential bits are ordered.

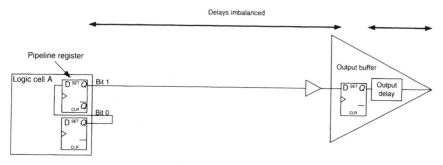

Figure 16.12   Performance reduction with register ordering.

The problem that arises with register ordering is that it may prevent a group of registers that have added pipelining for route balancing to be grouped together. This will prevent the additional registers from dividing the routing delay as originally intended. Consider the delay balancing register used above when packing registers into I/O buffers.

As can be seen from Figure 16.12, the benefit of the pipelined register has been eliminated due to the fact that it has been placed adjacent to its driver.

Register ordering should not be used when extra flip-flops have been added to divide routing delay.

16.9 PLACEMENT SEED

Designers typically do not like the concept that there may be a certain degree of randomness in their design flow or implementation tools. However, it is important to note that a placement for a given design with a set of constraints is not entirely deterministic. In other words, there is no single optimal placement for any given design; at least, not one that can be easily determined by today's technology prior to the placement process itself. In fact, as shown in Figure 16.13 this is not an obvious problem at all, even for an automated placer, to determine the optimal starting point for the placement process.

Thus, the placement tool needs to be "seeded" similar to a process for random number generation. The exact definition of a seed is a bit abstracted from the standpoint of the designer, but for useful purposes, different seeds essentially provide the placer with slightly different starting points by which the rest of the placement process branches. This is illustrated in Figure 16.14.

Some vendors extend this beyond simply the initialization to other "random" parameters of the placement process such as relative weighting of the various constraints, desirable placement distances, and estimates for routing congestion. Xilinx calls this the placement "cost table," which affects the various placement parameters, but abstracts this to an integer (1, 2, 3, etc.). Each integer corresponds with a different set of initial conditions and placement parameters, but because of the low level complexity, it is abstracted to a single number and the actual functionality is hidden from the parameter table.

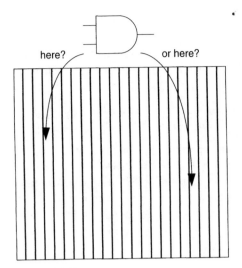

Figure 16.13   Initial placement.

The benefit to adjusting the placement seed is that the user can "take another shot" at meeting the timing constraints without actually changing anything in the design. This can be automated with a process called multipass place and route. When this option is selected, the place and route tool will run the implementation multiple times and automatically vary the seed on every run. This will of course require a large amount of time to produce effective results, so this is typically used on overnight runs where the target design is not already meeting the timing constraints in a repeatable manner.

It is important to note here that the place and route algorithms are actually very good regardless of the seed, and any changes to the various parameters will have a very small effect. The danger some newer designers run into is to spend

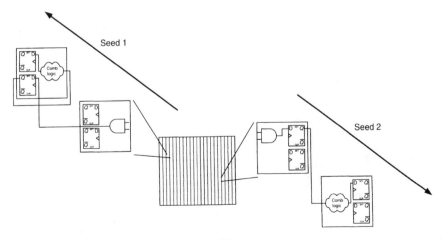

Figure 16.14   Different placement starting conditions.

large amounts of time focusing on this type of optimization to meet their timing constraints instead of focusing on the true issues in their design. Adjustment of the placer seed should only be used as a last resort when all architectural changes have been exhausted, all constraints have been added, and the timing violations are very small (hundreds of picoseconds or less).

Seed variances with multipass place and route should only be used as a last resort.

This should also be used only when repeatability is not a critical issue. If timing is so tight that tweaking the seed is required to meet timing, then any other changes to the RTL or constraints will more than likely lead to new random behavior in the design and require multiple runs of the placer with new variations in the seed. Thus, with this type of methodology, the designer will not have repeatability in the timing closure when small changes are made. This issue of repeatability is discussed in the following section.

16.10 GUIDED PLACE AND ROUTE

A very common situation for a designer is to spend days or weeks tweaking a design to meet timing, perhaps even using the automated seed variances as described above, only to be faced with a small change that could ripple through the existing placement and completely change the timing characteristics. At this point, the designer may have to reoptimize the design and possibly find a new seed that allows timing closure.

To avoid a situation where small changes initiate a domino effect on the placement of logic elements and subsequent change in overall performance characteristics (not to mention the run-time associated with reimplementing everything for a small change), FPGA place and route tools often provide a guide option that leverages the placement and routes from a prior implementation.

A placement guide will find all components that match the old implementation and lock the same elements to the same location. The advantage to this strategy is that the run time is dramatically reduced and the timing characteristics of the guided elements stay the same. Thus, a guide mode such as described above can provide the designer with a means to make small changes with a corresponding small change in the place and route process.

The place and route process following small changes should utilize a guide file to maximize consistency and minimize run time.

16.11 SUMMARY OF KEY POINTS

• A complete set of constraints should be created before any optimization takes place.

- Adjusting the voltage and temperature settings does not require any changes to the FPGA implementation and can provide an easy means to incrementally improve the worst-case performance.
- Placement effort has a dominant effect on performance, whereas routing effort has a relatively minor effect.
- Logic duplication should be used only on critical-path nets with multiple loads that cannot be physically localized.
- Optimization across hierarchy is not desirable when gate-level simulations are required.
- An extra pipeline register may be required for high-speed designs when packing registers into I/O.
- Setting a pack factor can help to determine the true utilization.
- Register ordering should not be used when extra flip-flops have been added to divide routing delay.
- Seed variances with multipass place and route should only be used as a last resort.
- The place and route process following small changes should utilize a guide file to maximize consistency and minimize run time.

Chapter 17

Example Design: Microprocessor

The simple RISC computer (SRC) implemented in this chapter is a widely used microprocessor model for academic purposes (a detailed description of the SRC architecture can be found in the book *Computer Systems Design and Architecture* by Vincent Heuring and Harry Jordan). The SRC microprocessor has a fairly generic architecture and is not complicated by many of the optimizations required in a commercial processor. Thus, it lends itself well to a straightforward pipelined implementation that helps to illustrate some of the various optimization strategies.

The objective of this chapter is to describe an implementation for the SRC processor and optimize the performance characteristics using various options described in previous chapters.

17.1 SRC ARCHITECTURE

The SRC is a 32-bit machine with 32 general-purpose registers (5-bit register addressing), 2^{32} bytes of main memory (addressable from any of the 32-bit registers), a single 32-bit program counter, and a 32-bit instruction register (Fig. 17.1).

The instruction classes are defined as follows:

- Load and store
- Branch
- Simple arithmetic: add, sub, invert
- Bitwise and shift

The pipelined implementation of the SRC processor partitions the main functional operations into distinct stages. Each stage is separated by a layer of

Advanced FPGA Design. By Steve Kilts
Copyright © 2007 John Wiley & Sons, Inc.

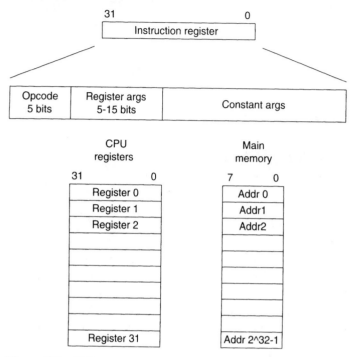

Figure 17.1 SRC registers and memory.

registers that limits the timing between the stages themselves. The pipeline is shown in Figure 17.2.

With the topology shown in Figure 17.2, it is possible to theoretically execute instructions such that one operation is completed every clock cycle. Note that because of a number of conditions that will periodically stall the pipeline, the number of clocks per instruction will be slightly larger than one, but for the purposes of this example we will assume that the fully pipelined implementation provides us with one instruction per clock cycle.

The code for the top level of the pipeline is shown in Appendix B. For brevity, the code for the submodules is not listed unless explicitly referred to for the sake of the example. There are a few interesting things to note about this implementation. First, it is assumed that the register file and system memory are located outside of the FPGA. These could be implemented inside the FPGA if block RAM resources are available, but if a large amount of system memory is required, it may be more cost effective to use an external memory device.

Second, note that each stage is separated by a layer of registers that define the pipeline. Any feedback signals as defined by the "Feedback" elements are also registers. Thus, any signal that passes to any stage is clearly defined as to the stage it belongs to and keeps the timing organized as the design becomes more complex.

Third, notice the parameterization of the registers themselves. The first parameter passed to the DRegister module is the width, which allows the register module to be reused among all DRegister elements.

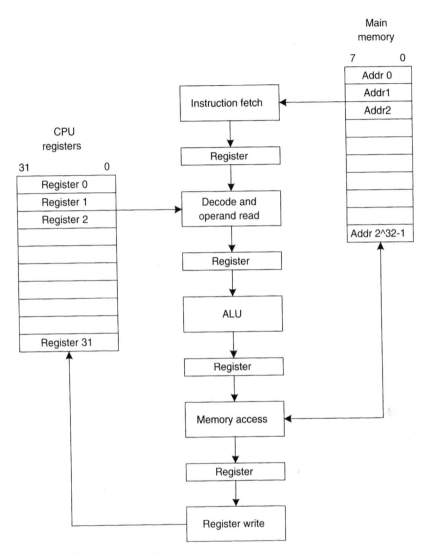

Figure 17.2 Main SRC pipeline.

Taking this pipelined design as a starting point, the following section describes some of the key optimization options and how they can be used to modify the performance of this design.

17.2 SYNTHESIS OPTIMIZATIONS

The SRC microprocessor was implemented in a Xilinx Virtex-2 part for this example as this is a mature FPGA technology that has been well characterized. A first-pass synthesis targeted at a XC2V250 with a 50-MHz clock gives us the performance as shown in Table 17.1.

Table 17.1   Initial Synthesis

Speed	63.6 MHz	Slack = 4.3 ns
LUTs	1386	45% utilization
Registers	353	11% utilization

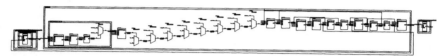

Figure 17.3   Serialized logic in critical path.

Note that due to the relatively slow timing constraint, the logic in the critical path is implemented in a serialized compact fashion and contains 23 levels of logic as shown in Figure 17.3.

17.2.1   Speed Versus Area

In the initial implementation, timing was easily met. This means that any paths that were not determined to be critical (worst-case delay) would be implemented with a compact architecture. This is not to say, however, that this is the maximum achievable frequency. If we push the timing constraint to 100 MHz, the synthesis tool will implement parallel architectures with a corresponding increase in area.

As can be seen from Table 17.2, the synthesis tool was able to expand the critical paths into higher speed implementations, thereby increasing the area 9% and the speed by more than 50%. Note that the timing increase does not display a 1 : 1 mapping to the area increases. This is because of the fact that the timing is only limited to the worst-case path. Therefore, the area increase is only an indicator that represents the number of paths that have been expanded to achieve the faster timing constraint. For example, the path that was previously the critical path with 23 levels of logic was remapped to a parallel structure as shown in Figure 17.4.

Assuming all reasonable architectural trade-offs have been made, there are various synthesis options that can be used to improve performance as described in the following sections.

Table 17.2   Increase in Target Frequency

Speed	97.6 MHz	Slack = −2.4 ns
LUTs	1538	50% utilization
Registers	358	11% utilization

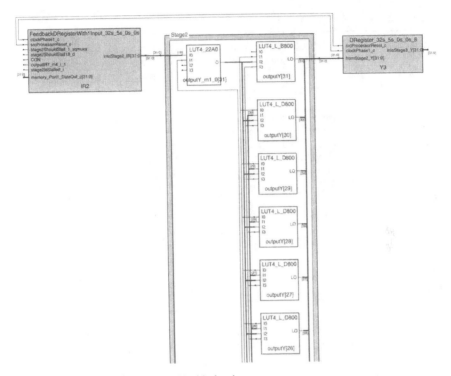

Figure 17.4   Parallel logic in the old critical path.

17.2.2 Pipelining

The new critical path now lies on the feedback path from IR2, through stage 2, and ending at X3 as shown in Figure 17.5.

To increase the performance of the new critical path, pipelining can be enabled to rebalance the logic around register layers. In this example, X3 is pushed into stage 2 and rebalanced around the multiplexing logic. Due to the replication of

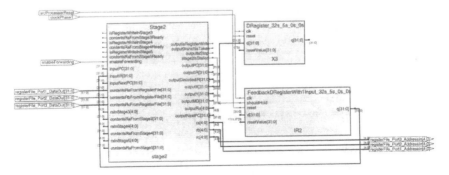

Figure 17.5   New critical path in stage 2.

Table 17.3   Results with Pipelining

Speed	100.7 MHz	Slack = 0.1 ns
LUTs	1554	50% utilization
Registers	678	22% utilization

Table 17.4   Results with Extra Pipelining

Speed	111.2 MHz	Slack = 0.1 ns
LUTs	1683	54% utilization
Registers	756	24% utilization

Table 17.5   Results with Physical Synthesis using Synplify Premiere

Speed	122 MHz	Slack = 0.2 ns
LUTs	1632	53% utilization
Registers	772	25% utilization

registers from the output of muxes to multiple inputs, the total register utilization essentially doubled with an insignificant change to the overall logic utilization.

The end result is that by pushing registers across logic boundaries and adding the necessary registers for all inputs, the critical path was reduced enough to meet timing at 100 MHz. Note that once the timing constraint was met, the synthesis tool stopped the pipelining operations to minimize overall register utilization. This is illustrated by the fact that bumping up the timing requirement to 110 MHz increases the additional register utilization as shown in Table 17.4.

17.2.3   Physical Synthesis

By engaging physical synthesis, we now begin to use physical placement information to optimize the performance of the design. With this placement information, we are able to target 120 MHz and easily close timing.

Note that the results in Table 17.5 are obtained by allowing the synthesis tool to determine all placement information with no user intervention. However, because this microprocessor design is highly pipelined, it is possible to guide the placement via a floorplan to achieve even better timing. This is discussed in the next section.

17.3   FLOORPLAN OPTIMIZATIONS

Because of the fact that the design is nicely partitioned into clearly defined pipeline stages, there is an opportunity to increase performance even further using predefined floorplan topologies.

17.3.1   Partitioned Floorplan

The first method that is used is that of a design partition. This is where the pipeline structure itself provides guidance as to the initial floorplan. In Figure 17.6, stages 2–4 are shown in block diagram form with intermediate register layers.

It is sensible to begin floorplanning with a similar physical layout. In the initial floorplan shown in Figure 17.7, the various pipeline stages are defined to be vertical sections of area to allow for ease of carry chain structures inside the FPGA.

For this floorplan, it was noted that the majority of the long paths reside between stages 1, 2, and 3. The three stages are given individual regions as shown in the floorplan of Figure 17.7. The data path flows from left to right, and a small amount of space is left between the stages for register placement. The critical paths between these modules are incrementally optimized as shown in Table 17.6. Note that the timing constraint has been increased to 125 MHz.

Our initial try did not produce results significantly better than the default physical synthesis approach. This is because of the large groups of constrained elements, the high aspect ratio of the physical regions themselves, and the fact that little importance was given to grouping only those paths that had tighter timing

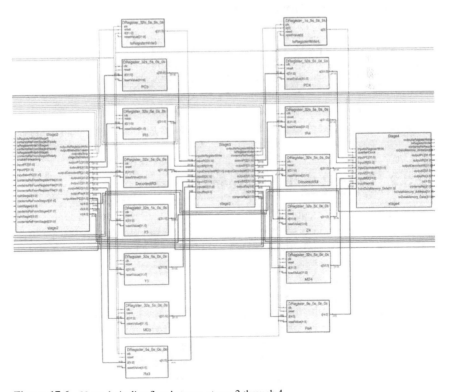

Figure 17.6   Natural pipeline flow between stages 2 through 4.

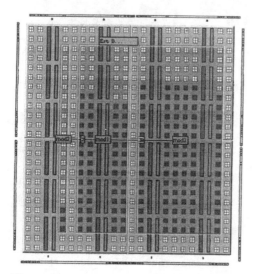

Figure 17.7   Floorplan that reflects pipeline flow.

Table 17.6   Results with Initial Floorplan

Speed	123 MHz	Slack = −0.17 ns	
LUTs	1840	59% utilization	
Registers	492	16% utilization	

requirements. To continue optimizing the performance, we must focus on the critical paths one at a time and physically group these together to minimize routing delays.

17.3.2   Critical-Path Floorplan: Abstraction 1

The iterative process begins by identifying the critical path and grouping these components together in a specified region. In this example, the critical paths lie between the registers surrounding stage 2. Figure 17.8 shows the worst-case path through stage 2.

With the first iterative method, we can first try constraining stage 2 and then adding in surrounding registers as required. The new floorplan is shown in Figure 17.9.

The advantage of the floorplan of Figure 17.9 is that it is more closely focused on the problem area and provides an even aspect ratio for ease of routing. The floorplan of Figure 17.9 represents stage 2 as well as the source register IR2 and the destination registers X3 and Y3. The performance is shown in Table 17.7.

As can be seen from these results, focusing our floorplan will allow us to achieve timing compliance at 125 MHz.

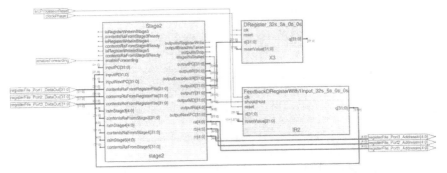

Figure 17.8   Worst case path through stage 2.

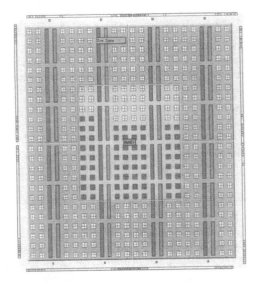

Figure 17.9   Floorplan with stage 2.

Table 17.7   Results with a Focused Floorplan

Speed	127.6 MHz	Slack = 0.16 ns
LUTs	1804	58% utilization
Registers	559	18% utilization

17.3.3   Critical-Path Floorplan: Abstraction 2

The final approach used to optimize the floorplan is to focus solely on the logic elements in the critical path itself. This has the advantage of constraining the difficult paths very tightly with minimal constraints on paths not under consideration. This is

Table 17.8 Results with Critical-Path Floorplanning

Speed	135 MHz	Slack = 0.0 ns
LUTs	1912	62% utilization
Registers	686	22% utilization

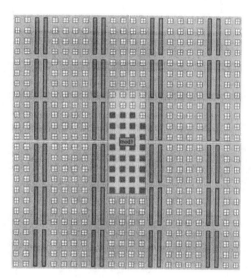

Figure 17.10 Critical path floorplan.

performed by taking the critical path from the default run and only constraining these components to a particular region to reduce all routing delays in the path.

In the default implementation, the critical path lies between stage 2 and the feedback register IR2. The implementation report indicates that approximately 50% of the total critical-path delay is routing delay only, which indicates there is room for improvement from critical-path floorplanning.

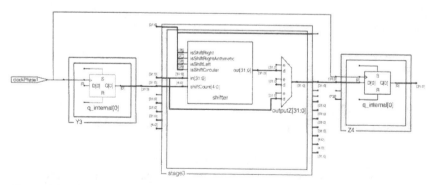

Figure 17.11 Final critical path.

The first two critical paths from our default run are from IR2 to Y3 and another between IR2 and X3. The elements from just these paths can be constrained to a region smaller than that of the previous floorplan.

By constraining only the critical paths to a smaller region, we again improve the performance. In Table 17.8, we are now targeting 135 MHz.

After these paths have been constrained to the region shown in Figure 17.10, the critical path now moves to the path between Y3 and Z4 through stage 3 as shown in Figure 17.11.

The routing delay has now dropped to 40% of the total delay (60% of the path is logic delay), which means that although there may be additional performance increases with further floorplanning, there will certainly be a diminishing rate of return relative to the effort involved.

Chapter 18

Static Timing Analysis

The goal of this chapter is to describe general static timing analysis (STA) as well as methods for performing static timing analysis on complex circuit configurations not commonly discussed in the context of STA such as asynchronous circuits, latches, and combinatorial feedback paths. Note that these latter structures are typically not recommended for FPGA designers mainly because of the difficulty in implementing them correctly. However, an advanced designer can use these structures if necessary as long as the associated issues (particularly the timing analysis) are understood.

During the course of this chapter, we will discuss the following topics:

- Summary of basic static timing analysis
- Understanding latches in STA
- Handling asynchronous circuits in STA with combinatorial logic or event-driven clocks

This chapter assumes the reader is already familiar with general static timing analysis but will provide a brief summary of the basic concepts.

18.1 STANDARD ANALYSIS

Static timing analysis, as it is referred to in this chapter, is the comprehensive analysis of all timing paths in a design relative to a set of constraints so as to determine whether a design is "timing compliant." The basic paths encountered by an FPGA designer will be input to flip-flop, flip-flop to flip-flop, and flip-flop to output as illustrated in Figure 18.1.

These have associated input delay, output delay, setup, and hold timing requirements. The setup timing analysis refers to the long-path analysis, and the hold timing refers to the short-path analysis. The maximum frequency is set by the longest path in the design, which is also referred to as the critical path. The

Advanced FPGA Design. By Steve Kilts
Copyright © 2007 John Wiley & Sons, Inc.

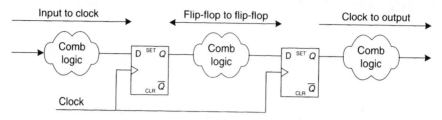

Figure 18.1 Typical synchronous timing paths.

path analysis takes into consideration more than just the logic delays as shown in Figure 18.2.

The maximum path delay is a function of the clock-to-Q delay, the logic and routing delays, the setup time, and the clock skew. The maximum frequency (minimum period) will be determined by the sum of the data-path delays minus the clock skew. In Chapter 1, we defined the maxium frequency in terms of all the associated delays, and here we define the minimum period as the inverse of this relationship as shown in Equation (18.1) (minimum clock period calculation): (again ignoring clock jitter).

Equation 18.1 minimum clock period calculation.

$$T_{min} = T_{clk\text{-}q} + T_{logic} + T_{routing} + T_{setup} - T_{skew} \qquad (18.1)$$

where T_{min} is minimum allowable period for clock, $T_{clk\text{-}q}$ is time from clock arrival until data arrives at Q, T_{logic} is propagation delay through logic between flip-flops, $T_{routing}$ is routing delay between flip-flops, T_{setup} is minimum time data must arrive at D before the next rising edge of clock (setup time), and T_{skew} is propagation delay of clock between the launch flip-flop and the capture flip-flop.

If the required period is greater than the minimum period defined by T_{min}, then there will be positive slack as shown in Figure 18.3.

As can be seen from the waveforms of Figure 18.3, positive slack occurs when the data arrive at the capture flip-flop before the capture clock less the setup time. If the data arrive after the capture clock less the setup time, the path will

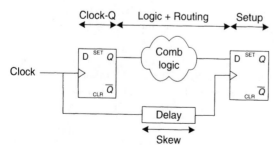

Figure 18.2 Elements of timing analysis.

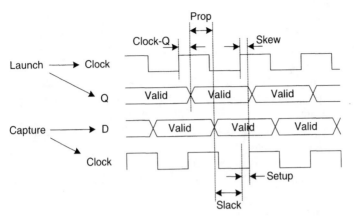

Figure 18.3   Positive slack.

have negative slack and will not meet timing. Setup analysis also applies to I/O. For an input, the launch clock-to-Q and external prop time are lumped into a single external delay for the analysis. Similarly, the output setup analysis will assume a single external prop time that includes the setup time and clock skew. Because these external delays are unknown to the FPGA timing analysis tool, they must be defined by the designer.

Hold violations occur when data arrives at a flip-flop too soon after the rising edge of the clock and are relatively easy to fix by adding additional buffering. Hold delay violations are rare in FPGA designs due to the built-in delay of the routing matrix. If a hold violation occurs, it usually indicates a clock skew problem.

In addition to defining accurate constraints for the system clocks, it is also important to define any path constraints that can be relaxed. The two most common constraints that fall into this category are multicycle and false paths. The multicycle path is illustrated in Figure 18.4.

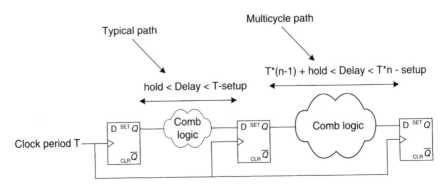

Figure 18.4   Multicycle path.

The multicycle path allows constrained signals n-cycles to propagate between timing end points. Note that even with a multicycle path, the setup and hold requirements are still in effect. This is because regardless of which clock edge the data arrives at, if it arrives too close to the clock edge, a metastable condition could still occur. A multicycle path may be added as a constraint if the data does not need to be used in the immediately following clock period. This could occur when interfacing to devices that do not require the immediate arrival of the data. This often occurs in DSP applications where there is a fixed sampling rate and an available number of clocks per sample.

A false path is similar to the multicycle path in that it is not required to propagate signals within a single clock period. The difference is that a false path is not logically possible as dictated by the design. In other words, even though the timing analysis tool sees a physical path from one point to the other through a series of logic gates, it is not logically possible for a signal to propagate between those two points during normal operation. This is illustrated in Figure 18.5.

Because the path will never be traversed during normal operation of the design, the static timing analysis tool ignores the path between these points and is not considered during timing optimization and critical-path analysis. The main difference between a multicycle path with many available cycles (large n) versus a false path is that the multicycle path will still be checked against setup and hold requirements and will still be included in the timing analysis. It is possible for a multicycle path to still fail timing, but a false path will never have any associated timing violations.

It is possible for a multicycle path to fail timing even with an arbitrarily high cycle constraint, but it is not possible for false paths to fail timing.

Once these constraints are in place, the STA tool can run a comprehensive analysis of every path in the design. There are a number of advantages of static timing analysis over simulation-based dynamic timing analysis as discussed in previous chapters:

· Dynamic timing analysis (simulations with timing information) will only catch a few problems. The utility of dynamic analysis is only as good as

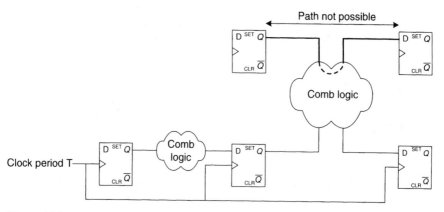

Figure 18.5   False path.

the person that wrote the simulation. STA, on the other hand, will catch all problems within the bounds of standard timing analysis. In other words, STA performs an exhaustive analysis on the design. The only requirement for the designer is to set up the necessary conditions and constraints.

- It is hard to model characteristics such as jitter, as this would create a very complex testbench and push the run times to unreasonable levels.

- With STA, combinations of various conditions can be verified with a proportional increase in run time. With dynamic timing analysis, combinations of various conditions can push the run time out exponentially.

- STA does not require any simulation cycles, and there is no overhead due to event schedulers.

- STA provides a larger scope of timing violations including positive and negative setup/hold, min and max transition, clock skew, glitch detection, and bus contention.

- STA tools have the ability to automatically detect and identify critical paths, violations, and asynchronous clocks.

The difficulty with STA arises when structures that are not well suited for STA are analyzed in this environment. Because of these constraints, dynamic timing analysis is often used to analyze these structures. The following sections discuss some of the more complex topologies in more detail and recommend methods to analyze these structures. Note that many of these structures are uncommon and are typically not recommended for FPGA designs. Listing these structures is not an endorsement to use them, but if an FPGA designer has good reason, then the suggestions can be used to analyze these successfully.

18.2 LATCHES

A latch can certainly be a useful device when creating very-high-speed chips such as microprocessors or with other high-speed pipelines. Latches should only be used by expert designers, and even then there must be a good justification to do so. Most of the time, latches are implemented unintentionally by poor coding styles and then show up in the STA reports to create confusion.

Latches are usually created unintentionally by poor coding styles.

One of the most common ways latches get implemented is with modules such as the one shown below.

```
// CAUTION: LATCH INFERENCE
module latch (
  output reg oDat,
  input      iClk, iDat);

  always @*
    if (iClk)  oDat <= iDat;
endmodule
```

In the above assignment, when iClk goes low the current value of oDat is held until the next assertion of iClk. Latches such as this are typically not desirable, but in some cases they are. Consider the following module where the latch was most likely induced unintentionally:

```
// POOR CODING STYLE
module latchinduced(
output reg oData,
input      iClk, iCtrl,
input      iData);
reg        rData;
reg        wData;

always @*
  if (iCtrl) wData <= rData;

always @(posedge iClk) begin
  rData          <= iData;
  oData          <= wData;
end
endmodule
```

In the above code segment, a latch is induced between the two D flip-flops as shown in Figure 18.6.

In this example, the data input and the data output are both registered with a standard D-type rising edge flip-flop, whereas the data is latched between the flip-flops. This circuit configuration is not one that is likely implemented intentionally, but it helps to cleanly illustrate the points regarding STA. We call this latch active-high because the data passes through when the control input is high, whereas the output is held when the control to the latch is low. From a timing analysis standpoint, we are not concerned with the condition where the control to the latch is high because the data is simply flowing through. Likewise, we are not concerned with the condition where the control is low because the output is frozen. What we are concerned with, however, is the timing relative to the point at which the data is actually latched; that is, during the transition from high to low.

When analyzing timing through a latch, the STA tool will be primarily concerned with the edge of the control signal that switches the latch into a hold state.

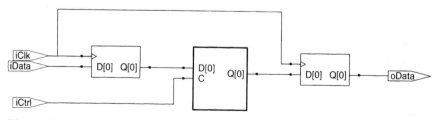

Figure 18.6   Induced latch.

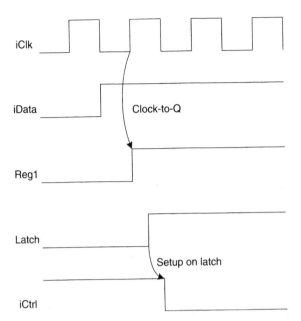

Figure 18.7   Timing compliance with induced latch.

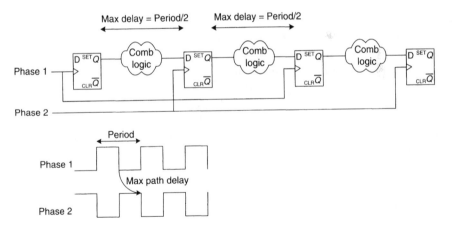

Figure 18.8   Dual-phase latching.

In a sense, the control to the latch is treated like a clock in STA, and the latch is treated like a falling-edge flip-flop. The waveforms shown in Figure 18.7 illustrate the conditions for timing compliance.

A common topology that utilizes latches is called two-phase latching. In this technique, one stage in a pipeline is latched with one polarity of the clock, while the stages on either side are latched with the opposite polarity as shown in Figure 18.8.

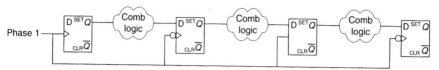

Figure 18.9 Alternating polarities with dual-phase latching.

A dual-phase module could be designed as shown in the example below.

```
// CAUTION: LATCH INFERENCE
module dualphase(
    output      oData,
    input       iCtrl, iNCtrl,
    input       iData);
    reg [3:0]   wData;

    assign oData        = wData[3];

    always @* begin
        if(iCtrl)   wData[0] <= iData;
        if(iNCtrl)  wData[1] <= wData[0];
        if(iCtrl)   wData[2] <= wData[1];
        if(iNCtrl)  wData[3] <= wData[2];
    end
endmodule
```

This would be implemented with a single clock signal and alternating latch polarities (Fig. 18.9).

In heavily pipelined designs such as microprocessors, a two-phase latching method can provide a smaller implementation (latches are smaller than flip-flops) and a lower power implementation (both edges of the clock are used for processing data). Note that this does not necessarily extend to all designs. If a circuit does not have a very well defined pipeline (and this is not always possible), then two-phase latching will be very difficult if not impossible.

> For heavily pipelined designs, a dual-phase latching topology can lead to a smaller and faster implementation.

The STA report for such an implementation will define a single clock domain; that is, the control signal to all of the latches. All of the paths will be reported between a rising edge (active low latch) and a falling edge (active high latch).

18.3 ASYNCHRONOUS CIRCUITS

In the context of this discussion, asynchronous circuits will refer to the broad class of circuits not within the domain of clocked element to clocked element paths that lend themselves to straightforward analysis. We will define a few of the most common configurations (although none are very common in general) and describe methods for analyzing the corresponding timing paths. Note that these are typically not recommended for FPGA designs, and so any designer that utilizes these must have a very good justification for doing so.

18.3.1 Combinatorial Feedback

A combinatorial feedback loop is a path through logic that begins at a net or wire, traverses combinatorial logic gates (AND, OR, MUX, etc.), and ends up at the same net without passing through any synchronous elements. Typically, a combinatorial feedback loop is a result of a design error and should be flagged as a warning by the synthesis tool. These can also cause problems for simulators (particularly if there are no combinatorial delays).

Combinatorial feedback typically indicates a coding error.

Depending on the specifics of the design, a combinatorial feedback loop will either exhibit oscillatory behavior (such as a free-running oscillator built with an inverter and a delay element) or contain properties of a memory element (basic latches and flip-flops can be built with cross-coupled NAND gates). Although it is possible to build a limited number of useful circuits with combinatorial feedback there is typically not an easy way to perform timing analysis because the end points are not sequential elements. Instead of measuring timing from one flip-flop to another, a designer needs to artificially define timing end points in the constraints. Take for example the free-running oscillator:

```
// BAD CODING STYLE.
module freeosc(
  output   oDat,
  input    iOscEn, iOutEn);
  wire     wdat;

  assign wdat = iOscEn ? !wdat: 0;
  assign oDat = iOutEn ? wdat : 0;
endmodule
```

This is implemented with a combinatorial loop as shown in Figure 18.10. This type of combinatorial feedback would certainly indicate a coding error, but it serves to illustrate the point.

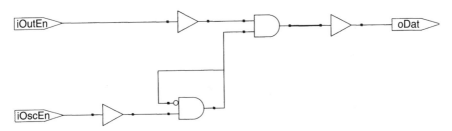

Figure 18.10 Accidental combinatorial feedback.

This design will synthesize unless the tool is explicitly prevented from synthesizing combinatorial loops. During analysis, the STA will only include paths from the input to the output. In fact, the STA tool will probably not even have the necessary data to measure the delay of the loop itself. To solve this problem, it is necessary to add a timing end point to at least one of the nets. In this case, we can add one to the net wdat (the net that is fed back). In a Xilinx implementation, the UCF (user constraints file) constraint would be:

```
NET         "wdat" TPSYNC = "looptime";
TIMESPEC    "ts_looptime" = FROM "looptime" TO "looptime"
                            1 ns;
```

In the above constraint, we place a timing end point on the net wdat as a part of the timing group looptime. We then place a constraint on looptime so the STA has something to measure against. The STA will now report about 850 ps as the loop delay.

18.4 SUMMARY OF KEY POINTS

- It is possible for a multicycle path to fail timing even with an arbitrarily high cycle constraint, but it is not possible for false paths to fail timing.
- Latches are usually created unintentionally by poor coding styles.
- When analyzing timing through a latch, the STA tool will be primarily concerned with the edge of the control signal that switches the latch into a hold state.
- For heavily pipelined designs, a dual-phase latching topology can lead to a smaller and faster implementation.
- Combinatorial feedback typically indicates a coding error.

Chapter 19

PCB Issues

The topic of PCB (Printed Circuit Board) design has been discussed in great detail in hundreds of other engineering texts. Thus, rather than regurgitate the same information that has been repeated dozens of times in these other books, I will refer the reader to one of these for a general discussion. There are a number of PCB design issues, however, that are specific (or exceptionally important) in an FPGA-based system.

During the course of this chapter, we will discuss the following topics:

- Characteristics of a proper FPGA power supply.
- Calculating, choosing, and placing decoupling capacitors.

19.1 POWER SUPPLY

The topic of power supply may seem trivial, but for FPGA applications in particular it is not. Poor use of decoupling around the FPGA can dramatically reduce the reliability of the FPGA, and worse yet, most problems will not be repeatable and may not show up in a lab environment at all (particularly if the nature of the failure is not understood). As mentioned in previous chapters, the worst kind of failure is one that is not repeatable.

19.1.1 Supply Requirements

Modern FPGAs that are fabricated with extremely small geometries will often have multiple supply voltages and complex power supply requirements. There have been many appnotes and white papers on the topic of power supply design, but as of yet there is no perfect power and decoupling solution for every FPGA application. The problem is that FPGAs themselves are so widely configurable relative to functionality, I/O, and system clock speeds that the corresponding

Advanced FPGA Design. By Steve Kilts
Copyright © 2007 John Wiley & Sons, Inc.

power requirements can be just as variable. Small, low-speed devices that reside in electrically shielded environments will not have a significant impact on the power rail, and excessive decoupling will only add unnecessary cost. Applications with high electrical noise or applications that produce excessive transients due to high-speed signals, on the other hand, will have a dramatic impact on the power supply, and a miscalculation of those effects can lead to a failure of the device.

The power supply requirements for FPGAs can be complex, but ignoring these requirements may lead to failures in the field that are not repeatable in the lab.

Overall requirements for FPGA power will vary from device to device, but out of these are a number of requirements that are worth noting in this discussion.

- Monotonicity
- Soft start
- Ramp control: min and max ramp times
- Peak-to-peak ripple
- Ripple rate of change
- Clean power for clock management
- Supply sequencing and tracking

Monotonicity is the requirement that during power-on, the rail is nondecreasing. That is, the power supply must always have a positive slope (or zero slope) and must never decrease (negative slope). For example, the time-domain power curve shown in Figure 19.1 illustrates a violation to this rule. In contrast, the curve in Figure 19.2 illustrates a monotonic power-on condition where the slope is always positive (or, more precisely, nondecreasing).

Soft start is a requirement that defines the amount of in-rush current that can be supplied to the FPGA during power-up. Most off-the-shelf power supply devices do not have built-in soft-start capability, and so an external circuit is often added to the PCB to satisfy this requirement.

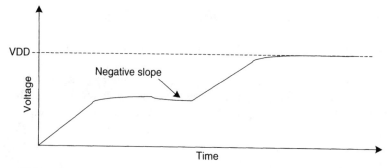

Figure 19.1   Nonmonotonic power curve.

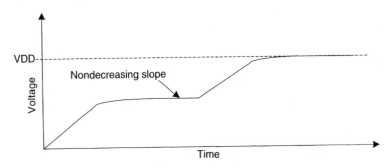

Figure 19.2   Monotonic power curve.

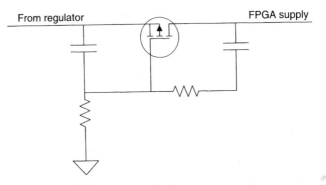

Figure 19.3   Typical soft-start circuit.

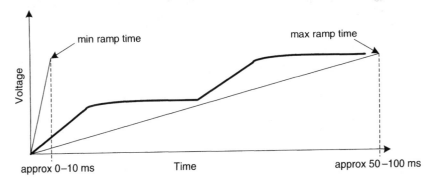

Figure 19.4   Min and max ramp times.

As can be seen from the circuit in Figure 19.3, if the power supply ramps up too quickly, the polarity of the gate on the pass-transistor will adjust itself to increase the output impedance of the power supply and reduce the rate of increase. This is directly related to the maximum ramp-time requirement.

The minimum and maximum ramp time requirements define the rate at which the power supply can increase during power-up. Ramping too quickly will cause

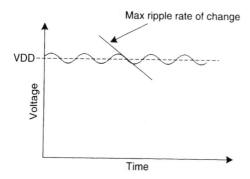

Figure 19.5 Ripple rate of change.

an in-rush condition as described above, and an unnecessarily slow ramp will dwell at threshold voltages and may not reset properly. This is illustrated in Figure 19.4.

For sensitive analog components such as clock-control circuitry, there is sometimes a requirement on the rate of change on the power supply ripple itself as illustrated in Figure 19.5. In other words, the power output must be clean of any high-frequency components above a certain threshold.

The requirement to have a clean supply for sensitive analog circuitry will typically require a linear supply on that particular rail to ensure that the majority of the frequency components have been removed (within the bandwidth of the linear supply itself).

In general, it is good design practice to add supply sequencing and supply tracking to a power supply design as shown in Figure 19.6. This comes from a basic principle that I/Os should not be powered before the core that drives the logic states. Most ICs including FPGAs will have circuitry built in to prevent any catastrophic failures due to unknown logic values driven to the outputs, but these problems have not always been successfully eliminated (despite what the data sheets say), and as a matter of good design practice, the core voltage should be powered before the I/O.

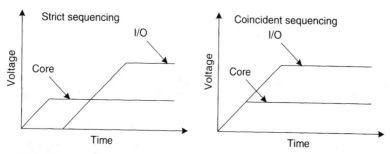

Figure 19.6 Supply sequencing.

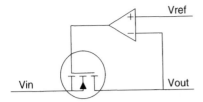

Figure 19.7   Linear regulation.

19.1.2 Regulation

The first and most obvious component in the overall power supply is the voltage regulator. A linear voltage regulator compensates for changes in the demand of current as shown in Figure 19.7.

If the load fluctuates momentarily, the current through the transistor will increase and drop the voltage at the output according to the series resistance of the transistor. The op-amp in the feedback senses this drop and increases the gate voltage, which has the effect of decreasing series resistance and increasing the output voltage to compensate for the drop. Although this feedback loop works well for relatively low- frequency signals in the low- to sub-megahertz range, the bandwidth of this type of loop is not fast enough to compensate for very-high-frequency transients.

19.2 DECOUPLING CAPACITORS

Decoupling capacitors are used to deliver small amounts of transient energy as the power rail fluctuates. Most PCB designers are trained to add capacitors to the power supply and place those capacitors close to the power pins of the ICs under consideration. The problem is that because this is not fully understood by many designers (particularly FPGA designers), the decoupling strategy is not executed properly, and capacitors are wasted without achieving any real benefit. For instance, many engineers will simply add one capacitor type to the power supply, and duplicate that for all power pins on the PCB. It is common to spread a number of 0.1-µF capacitors around the PCB with a bulk cap at the output of the regulator. Other PCB designers will use a variety of capacitor sizes, but not understanding why, they will use them in a proportion that is not appropriate for optimal decoupling.

Xilinx has published a great application note regarding power distribution system (PDS) design in XAPP623. It is recommended that the reader study this appnote for an in-depth discussion of this topic. The purpose of the following sections is to describe these issues as they apply to the FPGA designer.

19.2.1 Concept

Taking a step back, it is important to first understand the nature of a real capacitor. Every real-world capacitor will not only have capacitance as a dominant

Real capacitor

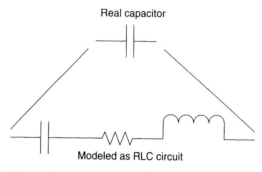

Modeled as RLC circuit

Figure 19.8 RLC capacitor model.

property but also inductance and resistance. A second-order model of a capacitor will be an RLC (Resistor-Inductor-Capacitor) circuit as shown in Figure 19.8.

Conceptually, the dominant impedance will be controlled by the capacitor at low frequencies and by the inductor at high frequencies. Due to this behavior, the real capacitor will have a high impedance at both very low and very high frequencies. There will also be a frequency at which the capacitive and inductive components effectively cancel each other out, and this will be a point of minimum impedance for the real capacitor modeled as a RLC circuit. This is shown in Figure 19.9.

This point of minimum impedance is called the resonance frequency and will define the band over which the decoupling cap will provide the most benefit for filtering out disturbances in the supply. Thus, it makes sense that to achieve a wide band of frequency attenuation, a relatively wide range of capacitors with different resonance frequencies will need to be used together.

As can be seen from Figure 19.10, by using a range of capacitors, we can achieve attenuation over a wide range of frequencies. Note, too, that to achieve significant attenuation with smaller capacitors (higher frequencies), a greater number of capacitors must be used. Because of the fact that a smaller capacitor

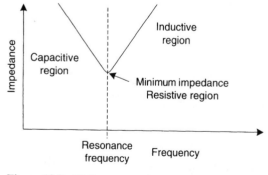

Figure 19.9 RLC resonance in a real capacitor.

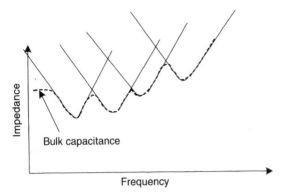

Figure 19.10   Wideband attenuation with multiple capacitors.

will hold a smaller amount of charge, a smaller capacitor will not be able to deliver as much energy to support the power rail as a larger capacitor.

A wide range of decoupling capacitors is required to attenuate a wide frequency range.

Although parasitics vary from technology to technology, as a general heuristic, a smaller package for an equivalent capacitance will tend to have smaller inductance. Thus, it is typically desirable to choose the smallest package available for a given capacitance to minimize the parasitic inductance and increase the bandwidth of the capacitor.

19.2.2 Calculating Values

Conceptually understanding capacitor resonance is one thing, but calculating the correct distribution of capacitors is another. Fortunately, there are a number of tools available to help automate this process. For instance, Xilinx provides a spreadsheet that allows a designer to plug in a range of capacitor values, and the corresponding impedance plot is graphed. This allows the designer to modify the capacitance values and visually adjust the attenuation plot for a specific set of criteria. As a general rule of thumb, capacitors of the following orders of magnitude should be used:

- 100–1000 μF
- 1–10 μF
- 0.1 μF
- 0.01 μF
- 0.001 μF

Additionally, for every reduction in the order of capacitance magnitude, the number of capacitors in that range should approximately double. In other words, if two 0.1-μF caps are used, four 0.01-μF caps should be used to achieve the same level of attenuation in the higher frequency band. The total number of caps, however, will be determined by the overall power requirements and the noise characteristics.

To make things even easier, Xilinx has added a feature to their XPower tool that calculates the recommended decoupling capacitors based not only on the static characteristics of the FPGA device but also on the dynamic power usage. For an example design and vector set (not shown here), XPower calculates 7 mW of dynamic power in the core and determines the following decoupling strategy.

```
Decoupling Network Summary:      Cap Range (µF)        #
-------------------------------------------------------------
Capacitor Recommendations:
Total for      Vccint :                                4
                              470.0  - 1000.0  :       1
                              0.0100 - 0.0470  :       1
                              0.0010 - 0.0047  :       2
                                     - - -
Total for      Vccaux :                                8
                              470.0  - 1000.0  :       1
                              0.0470 - 0.2200  :       1
                              0.0100 - 0.0470  :       2
                              0.0010 - 0.0047  :       4
                                     - - -
Total for      Vcco25 :                                8
                              470.0  - 1000.0  :       1
                              0.0470 - 0.2200  :       1
                              0.0100 - 0.0470  :       2
                              0.0010 - 0.0047  :       4
```

Minimally, a designer needs to put at least one capacitor near each power pin. Maximally, if high-frequency capacitors cannot fit near the FPGA power pins, their effectiveness is greatly reduced and may need to be removed. This is discussed in the next section.

19.2.3   Capacitor Placement

Aside from the actual capacitance values, poor placement of the cap can add to the parasitic inductance and diminish the utility of a decoupling cap as illustrated in Figure 19.11. Specifically, long narrow traces will tend to have a significant inductive component and will add to the effective inductance of the decoupling cap.

This increase in inductance will become a dominant feature in the RLC circuit for high frequencies, and as a consequence of this, it is extremely

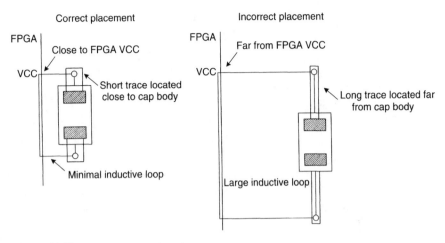

Figure 19.11   Decoupling capacitor placement.

important that the high-frequency caps be placed very close to the power pin of
the FPGA.

Higher-frequency caps should be placed closer to the FPGA power pins.

Because there will never be enough space to place all caps next to the power
pin on a real PCB layout, the trade-off must be made such that the smaller caps
are placed closer than the larger caps. The bulk caps (100–1000 μF), at the other
extreme, can be placed virtually anywhere on the PCB as they only respond to
slower transients.

At times, there is a temptation to share traces or vias as shown in Figure 19.12.
The effect of this is to increase the inductance and eliminate the overall effectiveness of the additional capacitors. It is recommended to assign a single via to each
capacitor pad and to connect the two through minimal trace lengths.

Minimize trace lengths to decoupling caps and assign a unique via to each
capacitor pad.

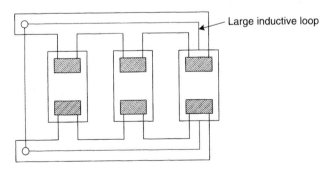

Figure 19.12   Poor design practice: shared vias.

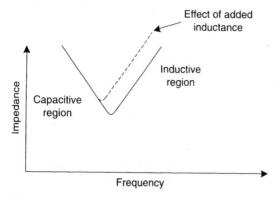

Figure 19.13 Bandwidth reduction with extra inductance.

If any additional parasitic inductance is added to the decoupling capacitor through any of the methods discussed here, the range of attenuation will be dramatically affected. As can be seen in Figure 19.13, the additional inductance will reduce the overall bandwidth of the capacitor and reduce the amount of attenuation at the resonance frequency.

For a more comprehensive study of PCB design, refer to the book High-Speed Digital Design by Howard Johnson and Martin Graham.

19.3 SUMMARY OF KEY POINTS

- The power supply requirements for FPGAs can be complex, but ignoring these requirements may lead to failures in the field that are not repeatable in the lab.

- A wide range of decoupling capacitors is required to attenuate a wide frequency range.

- Higher-frequency caps should be placed closer to the FPGA power pins.

- Minimize trace lengths to decoupling caps and assign a unique via to each capacitor pad.

Appendix A

Pipeline Stages for AES Encryption

The modules defined in this appendix belong to the AES core of Chapter 4.

```
// Provides necessary parameters to the
   AES implementation
// number of data words (always 32*4 = 128)
`define Nb 4

// 128 bit key mode
`define Nk4

// 192 bit key mode
//`define Nk6

// 256 bit key mode
//`define Nk8

`ifdef Nk4
   `define Nk 4
   `define Nr 10
`endif

`ifdef Nk6
   `define Nk 6
   `define Nr 12
`endif

`ifdef Nk8
   `define Nk 8
   `define Nr 14
`endif

// Performs the column mapping for MixColumns
module MapColumnEnc(
   output reg [31:0] oColumnOut,
   input             iClk, iReset,
```

Advanced FPGA Design. By Steve Kilts
Copyright © 2007 John Wiley & Sons, Inc.

```
input        [31:0] iColumnIn);
// intermediate Poly mult results
wire         [7:0]  S0x2, S1x2, S2x2, S3x2;
wire         [7:0]  S0x3, S1x3, S2x3, S3x3;
// Mapped cells in column
wire         [7:0]  S0PostMap, S1PostMap,
                    S2PostMap, S3PostMap;

// Modules that will perform poly mults over GF(2^8)
PolyMultx2Enc PolyMultS0x2(.iPolyIn(iColumnIn[31:24]),
                    .oPolyOut(S0x2));

PolyMultx2Enc PolyMultS1x2(.iPolyIn(iColumnIn[23:16]),
                    .oPolyOut(S1x2));

PolyMultx2Enc PolyMultS2x2(.iPolyIn(iColumnIn[15:8]),
                    .oPolyOut(S2x2));

PolyMultx2Enc PolyMultS3x2(.iPolyIn(iColumnIn[7:0]),
                    .oPolyOut(S3x2));

PolyMultx3Enc PolyMultS0x3(.iPolyIn(iColumnIn[31:24]),
                    .oPolyOut(S0x3));

PolyMultx3Enc PolyMultS1x3(.iPolyIn(iColumnIn[23:16]),
                    .oPolyOut(S1x3));

PolyMultx3Enc PolyMultS2x3(.iPolyIn(iColumnIn[15:8]),
                    .oPolyOut(S2x3));

PolyMultx3Enc PolyMultS3x3(.iPolyIn(iColumnIn[7:0]),
                    .oPolyOut(S3x3));

// Sum terms over GF(2)
assign S0PostMap = S0x2 ^ S1x3 ^ iColumnIn[15:8] ^
                   iColumnIn[7:0];
assign S1PostMap = iColumnIn[31:24] ^ S1x2 ^ S2x3 ^
                   iColumnIn[7:0];
assign S2PostMap = iColumnIn[31:24] ^
                   iColumnIn[23:16] ^ S2x2 ^ S3x3;
assign S3PostMap = S0x3 ^ iColumnIn[23:16] ^
                   iColumnIn[15:8] ^ S3x2;

always @(posedge iClk or negedge iReset) begin
  if (!iReset)
    oColumnOut <= 0;
  else // output is combination of post mapped cells
    oColumnOut <= {S0PostMap, S1PostMap, S2PostMap,
                   S3PostMap};
end
endmodule

module MixColumnsEnc(
  output reg [32 * 'Nb - 1:0] oBlockOut,
```

```verilog
output reg                      oValid,
input                           iClk, iReset,
input      [32 *'Nb - 1:0]      iBlockIn, // Data input to
                                be transformed
input                           iReady,
input      [3:0]                iRound);
reg        [32 *'Nb - 1:0]      BlockInHold; // registered
                                output
wire       [32 *'Nb - 1:0]      wPostMap;


MapColumnEnc MapColumnEnc0(.iClk(iClk),.iReset(iReset),
                           .iColumnIn({iBlockIn[127:120],
                                       iBlockIn[119:112],
                                       iBlockIn[111:104],
                                       iBlockIn[103:96]}),
                           .oColumnOut
                                  ({wPostMap[127:120],
                                    wPostMap[119:112],
                                    wPostMap[111:104],
                                    wPostMap[103:96]})));

MapColumnEnc MapColumnEnc1(.iClk(iClk), .iReset(iReset),
                           .iColumnIn({iBlockIn[95:88],
                                       iBlockIn[87:80],
                                       iBlockIn[79:72],
                                       iBlockIn[71:64]}),
                           .oColumnOut({wPostMap[95:88],
                                        wPostMap[87:80],
                                        wPostMap[79:72],
                                        wPostMap[71:64]})));

MapColumnEnc MapColumnEnc2(.iClk(iClk), .iReset(iReset),
                           .iColumnIn({iBlockIn[63:56],
                                       iBlockIn[55:48],
                                       iBlockIn[47:40],
                                       iBlockIn[39:32]}),
                           .oColumnOut({wPostMap[63:56],
                                        wPostMap[55:48],
                                        wPostMap[47:40],
                                        wPostMap[39:32]})));

MapColumnEnc MapColumnEnc3(.iClk(iClk), .iReset(iReset),
                           .iColumnIn({iBlockIn[31:24],
                                       iBlockIn[23:16],
                                       iBlockIn[15:8],
                                       iBlockIn[7:0]}),
                           .oColumnOut({wPostMap[31:24],
                                        wPostMap[23:16],
                                        wPostMap[15:8],
                                        wPostMap[7:0]})));
```

```
always @*
  if (iRound != 'Nr )
    oBlockOut   = wPostMap;
  else
    oBlockOut   = BlockInHold;

always @(posedge iClk or negedge iReset)
  if (!iReset) begin
    oValid      = 0;
    BlockInHold = 0;
  end
  else begin
    BlockInHold = iBlockIn;
    oValid      = iReady;
  end
endmodule

// Multiplies input poly by {02} over GF(2^8) and reduces
// mod m(x) = x^8 + x^4 + x^3 + x + 1
module PolyMultx2Enc(
  output [7:0]  oPolyOut,
  input  [7:0]  iPolyIn);
  wire   [8:0]  PolyPreShift, PolyPostShift, PolyReduced;

  assign PolyPreShift  = {1'b0, iPolyIn};
  assign PolyPostShift = PolyPreShift << 1;
  assign PolyReduced   = PolyPostShift[8]                  ?
                         (PolyPostShift ^ (9'b100011011)):
                         PolyPostShift;
  assign oPolyOut      = PolyReduced[7:0];
endmodule

// PolyMultx2Enc
// Multiplies input poly by {03} over GF(2^8) and reduces
// mod m(x) = x^8 + x^4 + x^3 + x + 1
module PolyMultx3Enc(
  output [7:0]  oPolyOut,
  input  [7:0]  iPolyIn);
  wire   [8:0]  PolyPreShift, PolyPostShift, PolyReduced;

  assign PolyPreShift  = {1'b0, iPolyIn};
  assign PolyPostShift = (PolyPreShift << 1) ^ PolyPreShift;
  assign PolyReduced   = PolyPostShift[8]                  ?
                         (PolyPostShift ^  (9'b100011011)):
                         PolyPostShift;
  assign oPolyOut      = PolyReduced[7:0];
endmodule

// PolyMultx3Enc
module ShiftRowsEnc(
```

```
output  [32 *'Nb - 1:0]  oBlockOut,
output                   oValid,
input   [32 *'Nb - 1:0]  iBlockIn,  // Data input to be
                                    transformed
input                    iReady);

assign oValid             = iReady;
assign oBlockOut[7:0]     = iBlockIn[39:32];
assign oBlockOut[15:8]    = iBlockIn[79:72];
assign oBlockOut[23:16]   = iBlockIn[119:112];
assign oBlockOut[31:24]   = iBlockIn[31:24];
assign oBlockOut[39:32]   = iBlockIn[71:64];
assign oBlockOut[47:40]   = iBlockIn[111:104];
assign oBlockOut[55:48]   = iBlockIn[23:16];
assign oBlockOut[63:56]   = iBlockIn[63:56];
assign oBlockOut[71:64]   = iBlockIn[103:96];
assign oBlockOut[79:72]   = iBlockIn[15:8];
assign oBlockOut[87:80]   = iBlockIn[55:48];
assign oBlockOut[95:88]   = iBlockIn[95:88];
assign oBlockOut[103:96]  = iBlockIn[7:0];
assign oBlockOut[111:104] = iBlockIn[47:40];
assign oBlockOut[119:112] = iBlockIn[87:80];
assign oBlockOut[127:120] = iBlockIn[127:120];
endmodule

// ShiftRowsEnc
// This block performs the SboxEnc transformation
   on the iBlockIn data
// and places it on oBlockOut
module SubBytesEnc(
   output reg [32 *'Nb - 1:0]  oBlockOut,
   output reg                  oValid,
   input                       iClk, iReset, iReady,
   input     [32 *'Nb - 1:0]   iBlockIn);  // Data input
                                           to be transformed
   wire      [32 *'Nb - 1:0]   wPostMap;

   SboxEnc SboxEnc1( .oPostMap(wPostMap[7:0]),
                     .iPreMap(iBlockIn[7:0]));
   SboxEnc SboxEnc2( .oPostMap(wPostMap[15:8]),
                     .iPreMap(iBlockIn[15:8]));
   SboxEnc SboxEnc3( .oPostMap(wPostMap[23:16]), .iPreMap(
                     iBlockIn[23:16]));
   SboxEnc SboxEnc4( .oPostMap(wPostMap[31:24]), .iPreMap(
                     iBlockIn[31:24]));
   SboxEnc SboxEnc5( .oPostMap(wPostMap[39:32]), .iPreMap(
                     iBlockIn[39:32]));
   SboxEnc SboxEnc6( .oPostMap(wPostMap[47:40]), .iPreMap(
                     iBlockIn[47:40]));
   SboxEnc SboxEnc7( .oPostMap(wPostMap[55:48]), .iPreMap(
                     iBlockIn[55:48]));
```

```
SboxEnc SboxEnc8 ( .oPostMap(wPostMap[63:56]), .iPreMap(
                    iBlockIn[63:56]));
SboxEnc SboxEnc9 ( .oPostMap(wPostMap[71:64]), .iPreMap(
                    iBlockIn[71:64]));
SboxEnc SboxEnc10(.oPostMap(wPostMap[79:72]), .iPreMap(
                    iBlockIn[79:72]));
SboxEnc SboxEnc11(.oPostMap(wPostMap[87:80]), .iPreMap(
                    iBlockIn[87:80]));
SboxEnc SboxEnc12(.oPostMap(wPostMap[95:88]), .iPreMap(
                    iBlockIn[95:88]));
SboxEnc SboxEnc13(.oPostMap(wPostMap[103:96]), .iPreMap(
                    iBlockIn[103:96]));
SboxEnc SboxEnc14(.oPostMap(wPostMap[111:104]), .iPreMap(
                    iBlockIn[111:104]));
SboxEnc SboxEnc15(.oPostMap(wPostMap[119:112]), .iPreMap(
                    iBlockIn[119:112]));
SboxEnc SboxEnc16(.oPostMap(wPostMap[127:120]), .iPreMap(
                    iBlockIn[127:120]));

always @(posedge iClk or negedge iReset)
  if (!iReset) begin
    oBlockOut <= 0;
    oValid    <= 0;
  end
  else begin
    oBlockOut  <= wPostMap;
    oValid     <= iReady;
  end
endmodule

// SubBytesEnc
// This block performs the AddRoundKey transformation
   on the iBlockIn data
// and places it on oBlockOut
module AddRoundKeyEnc(
    output reg [32 *'Nb - 1:0] oBlockOut,
    output reg                 oValid,
    input                      iClk, iReset, iReady,
    // Data input to be transformed
    input      [32 *'Nb - 1:0] iBlockIn, iRoundKey);
    reg        [32 *'Nb - 1:0] BlockOutStaged;
    reg                        ValidStaged;

  always @(posedge iClk or negedge iReset)
    if (!iReset) begin
      oBlockOut      <= 0;
      oValid         <= 0;
      BlockOutStaged <= 0;
      ValidStaged    <= 0;
    end
```

```verilog
    else begin
      BlockOutStaged <= iBlockIn ^ iRoundKey;
      ValidStaged    <= iReady;
      oBlockOut      <= BlockOutStaged;
      oValid         <= ValidStaged;
    end
  endmodule

// Registers all inputs
module InputRegsEnc(
  output  reg [32 *'Nk - 1:0] oKey,
  output  reg                 oReady, oKeysValid,
  output  reg [127:0]         oPlaintext,
  input                       iClk, iReset,
  input       [32 *'Nk - 1:0] iKey,
  input                       iNewKey, iReady,
  input       [127:0]         iPlaintext);
  reg         [32 *'Nk - 1:0] KeyReg;
  reg                         NewKeyReg, ReadyStaged;
  reg         [127:0]         PlaintextStaged;

  always @(posedge iClk or negedge iReset)
    if (!iReset) begin
      oKey           <= 0;
      oReady         <= 0;
      oPlaintext     <= 0;
      NewKeyReg      <= 0;
      KeyReg         <= 0;
      oKeysValid     <= 0;
      ReadyStaged    <= 0;
      PlaintextStaged <= 0;
    end
    else begin
      NewKeyReg      <= iNewKey;
      KeyReg         <= iKey;

      if (NewKeyReg) begin
        oKeysValid   <= 1;
        oKey         <= KeyReg;
      end
      else
        oKeysValid   <= 0;

      ReadyStaged    <= iReady;
      PlaintextStaged <= iPlaintext;
      oReady         <= ReadyStaged;
      oPlaintext     <= PlaintextStaged;
    end
  endmodule
```

```verilog
// RoundsIterEnc.v
// This module iterates the intermediate data through
   the round block
module RoundsIterEnc(
  output reg [32*'Nb-1:0] oBlockOut,
  output reg              oValid,
  input                   iClk, iReset,
  input      [32*'Nb-1:0] iBlockIn,
  input                   iReady,
  input      [127:0]      iRoundKey);
  reg        [3:0]        round;
  // keeps track of current  round
  reg                     ValidReg;
  reg        [127:0]      BlockOutReg;
  wire       [127:0]      wBlockIn, wBlockOut;
  wire                    wReady, wValid;

  assign wBlockIn  = iReady ? iBlockIn: wBlockOut;

  // ready is asserted when we have a new input or when the
  // previous round has completed and we are not done
  assign wReady    = iReady || (wValid && (round !=
  'Nr));


  RoundEnc Riter(.iClk(iClk), .iReset(iReset),
               .iBlockIn(wBlockIn), .iRoundKey(iRoundKey),
               .oBlockOut(wBlockOut), .iReady(wReady),
               .oValid(wValid), .iRound(round));

  always @(posedge iClk or negedge iReset)
    if(!iReset) begin
      round         <= 0;
      oValid        <= 0;
      oBlockOut     <= 0;
      ValidReg      <= 0;
      BlockOutReg   <= 0;
    end
    else begin
      oValid        <= ValidReg;
      oBlockOut     <= BlockOutReg;

    if(iReady) begin
      round         <= 1;
      ValidReg      <= 0;
  end
  else if(wValid && (round != 0)) begin
    // rounds continue and data has completed another round
    if(round == 'Nr) begin
      // data has completed last round
      round         <= 0;
```

```verilog
      ValidReg       <= 1;
      BlockOutReg    <= wBlockOut;
   end
   else begin
      // data will continue through rounds
      round          <= round + 1;
      ValidReg       <= 0;
   end
   end
      else ValidReg <= 0;
   end
   endmodule

// SboxEnc.v
// returns mapped value from LUT
module SboxEnc(
   output reg [7:0] oPostMap,
   input      [7:0] iPreMap);

   // Define the Sbox
   always @*
   case(iPreMap[7:0])
      8'h00: oPostMap = 8'h63;
      8'h01: oPostMap = 8'h7c;
      8'h02: oPostMap = 8'h77;
      8'h03: oPostMap = 8'h7b;
      8'h04: oPostMap = 8'hf2;
      8'h05: oPostMap = 8'h6b;
      8'h06: oPostMap = 8'h6f;
      8'h07: oPostMap = 8'hc5;
      8'h08: oPostMap = 8'h30;
      8'h09: oPostMap = 8'h01;
      8'h0a: oPostMap = 8'h67;
      8'h0b: oPostMap = 8'h2b;
      8'h0c: oPostMap = 8'hfe;
      8'h0d: oPostMap = 8'hd7;
      8'h0e: oPostMap = 8'hab;
      8'h0f: oPostMap = 8'h76;
      8'h10: oPostMap = 8'hca;
      8'h11: oPostMap = 8'h82;
      8'h12: oPostMap = 8'hc9;
      8'h13: oPostMap = 8'h7d;
      8'h14: oPostMap = 8'hfa;
      8'h15: oPostMap = 8'h59;
      8'h16: oPostMap = 8'h47;
      8'h17: oPostMap = 8'hf0;
      8'h18: oPostMap = 8'had;
      8'h19: oPostMap = 8'hd4;
      8'h1a: oPostMap = 8'ha2;
```

```
8'h1b: oPostMap = 8'haf;
8'h1c: oPostMap = 8'h9c;
8'h1d: oPostMap = 8'ha4;
8'h1e: oPostMap = 8'h72;
8'h1f: oPostMap = 8'hc0;
8'h20: oPostMap = 8'hb7;
8'h21: oPostMap = 8'hfd;
8'h22: oPostMap = 8'h93;
8'h23: oPostMap = 8'h26;
8'h24: oPostMap = 8'h36;
8'h25: oPostMap = 8'h3f;
8'h26: oPostMap = 8'hf7;
8'h27: oPostMap = 8'hcc;
8'h28: oPostMap = 8'h34;
8'h29: oPostMap = 8'ha5;
8'h2a: oPostMap = 8'he5;
8'h2b: oPostMap = 8'hf1;
8'h2c: oPostMap = 8'h71;
8'h2d: oPostMap = 8'hd8;
8'h2e: oPostMap = 8'h31;
8'h2f: oPostMap = 8'h15;
8'h30: oPostMap = 8'h04;
8'h31: oPostMap = 8'hc7;
8'h32: oPostMap = 8'h23;
8'h33: oPostMap = 8'hc3;
8'h34: oPostMap = 8'h18;
8'h35: oPostMap = 8'h96;
8'h36: oPostMap = 8'h05;
8'h37: oPostMap = 8'h9a;
8'h38: oPostMap = 8'h07;
8'h39: oPostMap = 8'h12;
8'h3a: oPostMap = 8'h80;
8'h3b: oPostMap = 8'he2;
8'h3c: oPostMap = 8'heb;
8'h3d: oPostMap = 8'h27;
8'h3e: oPostMap = 8'hb2;
8'h3f: oPostMap = 8'h75;
8'h40: oPostMap = 8'h09;
8'h41: oPostMap = 8'h83;
8'h42: oPostMap = 8'h2c;
8'h43: oPostMap = 8'h1a;
8'h44: oPostMap = 8'h1b;
8'h45: oPostMap = 8'h6e;
8'h46: oPostMap = 8'h5a;
8'h47: oPostMap = 8'ha0;
8'h48: oPostMap = 8'h52;
8'h49: oPostMap = 8'h3b;
8'h4a: oPostMap = 8'hd6;
8'h4b: oPostMap = 8'hb3;
```

```
8'h4c:  oPostMap = 8'h29;
8'h4d:  oPostMap = 8'he3;
8'h4e:  oPostMap = 8'h2f;
8'h4f:  oPostMap = 8'h84;
8'h50:  oPostMap = 8'h53;
8'h51:  oPostMap = 8'hd1;
8'h52:  oPostMap = 8'h00;
8'h53:  oPostMap = 8'hed;
8'h54:  oPostMap = 8'h20;
8'h55:  oPostMap = 8'hfc;
8'h56:  oPostMap = 8'hb1;
8'h57:  oPostMap = 8'h5b;
8'h58:  oPostMap = 8'h6a;
8'h59:  oPostMap = 8'hcb;
8'h5a:  oPostMap = 8'hbe;
8'h5b:  oPostMap = 8'h39;
8'h5c:  oPostMap = 8'h4a;
8'h5d:  oPostMap = 8'h4c;
8'h5e:  oPostMap = 8'h58;
8'h5f:  oPostMap = 8'hcf;
8'h60:  oPostMap = 8'hd0;
8'h61:  oPostMap = 8'hef;
8'h62:  oPostMap = 8'haa;
8'h63:  oPostMap = 8'hfb;
8'h64:  oPostMap = 8'h43;
8'h65:  oPostMap = 8'h4d;
8'h66:  oPostMap = 8'h33;
8'h67:  oPostMap = 8'h85;
8'h68:  oPostMap = 8'h45;
8'h69:  oPostMap = 8'hf9;
8'h6a:  oPostMap = 8'h02;
8'h6b:  oPostMap = 8'h7f;
8'h6c:  oPostMap = 8'h50;
8'h6d:  oPostMap = 8'h3c;
8'h6e:  oPostMap = 8'h9f;
8'h6f:  oPostMap = 8'ha8;
8'h70:  oPostMap = 8'h51;
8'h71:  oPostMap = 8'ha3;
8'h72:  oPostMap = 8'h40;
8'h73:  oPostMap = 8'h8f;
8'h74:  oPostMap = 8'h92;
8'h75:  oPostMap = 8'h9d;
8'h76:  oPostMap = 8'h38;
8'h77:  oPostMap = 8'hf5;
8'h78:  oPostMap = 8'hbc;
8'h79:  oPostMap = 8'hb6;
8'h7a:  oPostMap = 8'hda;
8'h7b:  oPostMap = 8'h21;
8'h7c:  oPostMap = 8'h10;
```

```
8'h7d: oPostMap = 8'hff;
8'h7e: oPostMap = 8'hf3;
8'h7f: oPostMap = 8'hd2;
8'h80: oPostMap = 8'hcd;
8'h81: oPostMap = 8'h0c;
8'h82: oPostMap = 8'h13;
8'h83: oPostMap = 8'hec;
8'h84: oPostMap = 8'h5f;
8'h85: oPostMap = 8'h97;
8'h86: oPostMap = 8'h44;
8'h87: oPostMap = 8'h17;
8'h88: oPostMap = 8'hc4;
8'h89: oPostMap = 8'ha7;
8'h8a: oPostMap = 8'h7e;
8'h8b: oPostMap = 8'h3d;
8'h8c: oPostMap = 8'h64;
8'h8d: oPostMap = 8'h5d;
8'h8e: oPostMap = 8'h19;
8'h8f: oPostMap = 8'h73;
8'h90: oPostMap = 8'h60;
8'h91: oPostMap = 8'h81;
8'h92: oPostMap = 8'h4f;
8'h93: oPostMap = 8'hdc;
8'h94: oPostMap = 8'h22;
8'h95: oPostMap = 8'h2a;
8'h96: oPostMap = 8'h90;
8'h97: oPostMap = 8'h88;
8'h98: oPostMap = 8'h46;
8'h99: oPostMap = 8'hee;
8'h9a: oPostMap = 8'hb8;
8'h9b: oPostMap = 8'h14;
8'h9c: oPostMap = 8'hde;
8'h9d: oPostMap = 8'h5e;
8'h9e: oPostMap = 8'h0b;
8'h9f: oPostMap = 8'hdb;
8'ha0: oPostMap = 8'he0;
8'ha1: oPostMap = 8'h32;
8'ha2: oPostMap = 8'h3a;
8'ha3: oPostMap = 8'h0a;
8'ha4: oPostMap = 8'h49;
8'ha5: oPostMap = 8'h06;
8'ha6: oPostMap = 8'h24;
8'ha7: oPostMap = 8'h5c;
8'ha8: oPostMap = 8'hc2;
8'ha9: oPostMap = 8'hd3;
8'haa: oPostMap = 8'hac;
8'hab: oPostMap = 8'h62;
8'hac: oPostMap = 8'h91;
8'had: oPostMap = 8'h95;
```

```
8'hae: oPostMap = 8'he4;
8'haf: oPostMap = 8'h79;
8'hb0: oPostMap = 8'he7;
8'hb1: oPostMap = 8'hc8;
8'hb2: oPostMap = 8'h37;
8'hb3: oPostMap = 8'h6d;
8'hb4: oPostMap = 8'h8d;
8'hb5: oPostMap = 8'hd5;
8'hb6: oPostMap = 8'h4e;
8'hb7: oPostMap = 8'ha9;
8'hb8: oPostMap = 8'h6c;
8'hb9: oPostMap = 8'h56;
8'hba: oPostMap = 8'hf4;
8'hbb: oPostMap = 8'hea;
8'hbc: oPostMap = 8'h65;
8'hbd: oPostMap = 8'h7a;
8'hbe: oPostMap = 8'hae;
8'hbf: oPostMap = 8'h08;
8'hc0: oPostMap = 8'hba;
8'hc1: oPostMap = 8'h78;
8'hc2: oPostMap = 8'h25;
8'hc3: oPostMap = 8'h2e;
8'hc4: oPostMap = 8'h1c;
8'hc5: oPostMap = 8'ha6;
8'hc6: oPostMap = 8'hb4;
8'hc7: oPostMap = 8'hc6;
8'hc8: oPostMap = 8'he8;
8'hc9: oPostMap = 8'hdd;
8'hca: oPostMap = 8'h74;
8'hcb: oPostMap = 8'h1f;
8'hcc: oPostMap = 8'h4b;
8'hcd: oPostMap = 8'hbd;
8'hce: oPostMap = 8'h8b;
8'hcf: oPostMap = 8'h8a;
8'hd0: oPostMap = 8'h70;
8'hd1: oPostMap = 8'h3e;
8'hd2: oPostMap = 8'hb5;
8'hd3: oPostMap = 8'h66;
8'hd4: oPostMap = 8'h48;
8'hd5: oPostMap = 8'h03;
8'hd6: oPostMap = 8'hf6;
8'hd7: oPostMap = 8'h0e;
8'hd8: oPostMap = 8'h61;
8'hd9: oPostMap = 8'h35;
8'hda: oPostMap = 8'h57;
8'hdb: oPostMap = 8'hb9;
8'hdc: oPostMap = 8'h86;
8'hdd: oPostMap = 8'hc1;
8'hde: oPostMap = 8'h1d;
```

```
      8'hdf:  oPostMap = 8'h9e;
      8'he0:  oPostMap = 8'he1;
      8'he1:  oPostMap = 8'hf8;
      8'he2:  oPostMap = 8'h98;
      8'he3:  oPostMap = 8'h11;
      8'he4:  oPostMap = 8'h69;
      8'he5:  oPostMap = 8'hd9;
      8'he6:  oPostMap = 8'h8e;
      8'he7:  oPostMap = 8'h94;
      8'he8:  oPostMap = 8'h9b;
      8'he9:  oPostMap = 8'h1e;
      8'hea:  oPostMap = 8'h87;
      8'heb:  oPostMap = 8'he9;
      8'hec:  oPostMap = 8'hce;
      8'hed:  oPostMap = 8'h55;
      8'hee:  oPostMap = 8'h28;
      8'hef:  oPostMap = 8'hdf;
      8'hf0:  oPostMap = 8'h8c;
      8'hf1:  oPostMap = 8'ha1;
      8'hf2:  oPostMap = 8'h89;
      8'hf3:  oPostMap = 8'h0d;
      8'hf4:  oPostMap = 8'hbf;
      8'hf5:  oPostMap = 8'he6;
      8'hf6:  oPostMap = 8'h42;
      8'hf7:  oPostMap = 8'h68;
      8'hf8:  oPostMap = 8'h41;
      8'hf9:  oPostMap = 8'h99;
      8'hfa:  oPostMap = 8'h2d;
      8'hfb:  oPostMap = 8'h0f;
      8'hfc:  oPostMap = 8'hb0;
      8'hfd:  oPostMap = 8'h54;
      8'hfe:  oPostMap = 8'hbb;
      8'hff:  oPostMap = 8'h16;
   endcase
endmodule
```

Appendix B

Top-Level Module for the SRC Processor

The module defined in this appendix belongs to the SRC processor example of Chapter 17.

```
module SrcProcessor(
    output          hasExecutedStop,
    output [31:0]   memory_Port1_DataIn,
    output [31:0]   memory_Port1_AddressIn,
    output          memory_Port1_WriteStrobe,
    output [31:0]   memory_Port2_DataIn,
    output [31:0]   memory_Port2_AddressIn,
    output          memory_Port2_WriteStrobe,
    output [31:0]   registerFile_Port1_DataIn,
    output [4:0]    registerFile_Port1_AddressIn,
    output          registerFile_Port1_WriteStrobe,
    output [31:0]   registerFile_Port2_DataIn,
    output [4:0]    registerFile_Port2_AddressIn,
    output          registerFile_Port2_WriteStrobe,
    output [31:0]   registerFile_Port3_DataIn,
    output [4:0]    registerFile_Port3_AddressIn,
    output          registerFile_Port3_WriteStrobe,
    output [31:0]   registerFile_Port4_DataIn,
    output [4:0]    registerFile_Port4_AddressIn,
    output          registerFile_Port4_WriteStrobe,
    input           clock,
    input           srcProcessorReset,
    input           canRun,
    input [31:0]    memory_Port1_DataOut,
    input [31:0]    memory_Port2_DataOut,
    input [31:0]    registerFile_Port1_DataOut,
    input [31:0]    registerFile_Port2_DataOut,
    input [31:0]    registerFile_Port3_DataOut,
    input [31:0]    registerFile_Port4_DataOut,
```

Advanced FPGA Design. By Steve Kilts
Copyright © 2007 John Wiley & Sons, Inc.

```
input          enableForwarding,
input  [31:0]  cycleNumber);
wire           hasDecodedStop;
wire           hasExecutedStop;

// Stage1 declarations
wire   [31:0]  intoStage1_PC;
wire   [31:0]  intoStage1_IR;
wire           stage2IsStalled;
wire           intoStage1_CanRun;
wire           intoStage1_ShouldStop;
wire           fromStage2_BranchIsTaken;
wire   [31:0]  fromStage1_PC;
wire   [31:0]  fromStage1_IR;
wire   [31:0]  fromStage1_NextPC;
wire           stage1IsStalled;

// Stage2 declarations
wire   [31:0]  intoStage2_PC;
wire   [31:0]  intoStage2_IR;
wire   [31:0]  intoStage2_NextPC;
wire   [31:0]  fromStage2_PC;
wire   [31:0]  fromStage2_IR;
wire   [31:0]  fromStage2_DecodedIR;
wire   [31:0]  fromStage2_X;
wire   [31:0]  fromStage2_Y;
wire   [31:0]  fromStage2_MD;
wire           fromStage2_IsRegisterWrite;
wire   [4:0]   fromStage2_Ra;
wire   [31:0]  fromStage2_NextPC;
wire           fromStage2_IsStop;
wire   [4:0]   ra;
wire   [31:0]  contentsRaFromRegisterFile;
wire   [4:0]   rb;
wire   [31:0]  contentsRbFromRegisterFile;
wire   [4:0]   rc;
wire   [31:0]  contentsRcFromRegisterFile;
wire           isRegisterWriteInStage3;
wire   [4:0]   raInStage3;
wire   [31:0]  contentsRaFromStage3;
wire           contentsRaFromStage3Ready;
wire           isRegisterWriteInStage4;
wire   [4:0]   raInStage4;
wire   [31:0]  contentsRaFromStage4;
wire           contentsRaFromStage4Ready;
wire           isRegisterWriteInStage5;
wire   [4:0]   raInStage5;
wire   [31:0]  contentsRaFromStage5;
wire           contentsRaFromStage5Ready;
wire           enableForwarding;
```

```
// Stage3 declarations
wire [31:0] intoStage3_PC;
wire [31:0] intoStage3_IR;
wire [31:0] intoStage3_DecodedIR;
wire [31:0] intoStage3_X;
wire [31:0] intoStage3_Y;
wire [31:0] intoStage3_MD;
wire        intoStage3_IsRegisterWrite;
wire [4:0]  intoStage3_Ra;
wire [31:0] fromStage3_PC;
wire [31:0] fromStage3_IR;
wire [31:0] fromStage3_DecodedIR;
wire [31:0] fromStage3_Z;
wire [31:0] fromStage3_MD;
wire        fromStage3_IsRegisterWrite;
wire [4:0]  fromStage3_Ra;

// Stage4 declarations
wire [31:0] intoStage4_PC;
wire [31:0] intoStage4_IR;
wire [31:0] intoStage4_DecodedIR;
wire [31:0] intoStage4_Z;
wire [31:0] intoStage4_MD;
wire        intoStage4_IsRegisterWrite;
wire [4:0]  intoStage4_Ra;
wire [31:0] fromStage4_PC;
wire [31:0] fromStage4_IR;
wire [31:0] fromStage4_DecodedIR;
wire [31:0] fromStage4_Z;
wire        fromStage4_IsRegisterWrite;
wire [4:0]  fromStage4_Ra;
wire [31:0] toDataMemory_Address;
wire [31:0] toDataMemory_Data;
wire        toDataMemory_WriteStrobe;
wire [31:0] fromDataMemory_Data;

// Stage5 declarations
wire [31:0] intoStage5_PC;
wire [31:0] intoStage5_IR;
wire [31:0] intoStage5_DecodedIR;
wire [31:0] intoStage5_Z;
wire        intoStage5_IsRegisterWrite;
wire [4:0]  intoStage5_Ra;
wire        fromStage5_IsStop;
wire [4:0]  toRegisterFile_Address;
wire [31:0] toRegisterFile_Data;
wire        toRegisterFile_WriteStrobe;

// unused but included here for completeness
wire [31:0]  fromRegisterFile_Data;
```

```
// logic for interface to instruction and data memory
assign  intoStage1_IR                  = memory_Port1_
                                         DataOut;
assign  memory_Port1_DataIn            = 32'b0;
assign  memory_Port1_AddressIn         = intoStage1_PC;
assign  memory_Port1_WriteStrobe       = 1'b0;

assign  fromDataMemory_Data            = memory_Port2_
                                         DataOut;
assign  memory_Port2_DataIn            = toDataMemory_
                                         Data;
assign  memory_Port2_AddressIn         = toDataMemory_
                                         Address;
assign  memory_Port2_WriteStrobe       = toDataMemory_
                                         WriteStrobe;

// logic for interface to register file
assign  contentsRaFromRegisterFile     = registerFile_
                                         Port1_DataOut;
assign  registerFile_Port1_DataIn      = 32'b0;
assign  registerFile_Port1_AddressIn   = ra;
assign  registerFile_Port1_WriteStrobe = 1'b0;

assign  contentsRbFromRegisterFile     = registerFile_
                                         Port2_DataOut;
assign  registerFile_Port2_DataIn      = 32'b0;
assign  registerFile_Port2_AddressIn   = rb;
assign  registerFile_Port2_WriteStrobe = 1'b0;

assign  contentsRcFromRegisterFile     = registerFile_
                                         Port3_DataOut;
assign  registerFile_Port3_DataIn      = 32'b0;
assign  registerFile_Port3_AddressIn   = rc;
assign  registerFile_Port3_WriteStrobe = 1'b0;

assign  fromRegisterFile_Data          = registerFile_
                                         Port4_DataOut;
assign  registerFile_Port4_DataIn      = toRegisterFile_
                                         Data;
assign  registerFile_Port4_AddressIn   = toRegisterFile_
                                         Address;
assign  registerFile_Port4_WriteStrobe = toRegisterFile_
                                         WriteStrobe;

//
// Module: FeedbackDRegisterWith2Inputs
//
// Description:
//    Special register for PC
```

```
//
// Inputs:
//    clk                     <-- clock
//    shouldHold              <-- stage1IsStalled
//    d0                      <-- fromStage1_NextPC
//    d1                      <-- fromStage2_NextPC
//    select                  <-- fromStage2_BranchIsTaken
//    reset                   <-- srcProcessorReset
//    resetValue
//
// Outputs:
//    q                       --> intoStage1_PC
//


FeedbackDRegisterWith2Inputs #(32, 5, 0, 0) PC
                            ( clock,
                              stage1IsStalled,
                              fromStage1_NextPC,
                              fromStage2_NextPC,
                              fromStage2_BranchIsTaken,
                              intoStage1_PC,
                              srcProcessorReset,
                              32'b0);


or IntoStage1_ShouldStop (        intoStage1_ShouldStop,
                                  hasDecodedStop,
                                  fromStage2_IsStop);


assign    intoStage1_CanRun  = canRun;


//
// Module: Stage1
//
// Description:
//    Instruction Fetch
//
// Inputs:
//    inputPC                 <-- intoStage1_PC
//    inputIR                 <-- intoStage1_IR
//    stage2IsStalled         <-- stage2IsStalled
//    canRun                  <-- intoStage1_CanRun
//    shouldStop              <-- intoStage1_ShouldStop
//    branchIsTakenInStage2   <-- fromStage2_BranchIsTaken
//
// Outputs:
//    outputPC                --> fromStage1_PC
//    outputIR                --> fromStage1_IR
//    outputNextPC            --> fromStage1_NextPC
```

```
//      stage1IsStalled          --> stage1IsStalled
//

Stage1 stage1 (intoStage1_PC,
               intoStage1_IR,
               stage2IsStalled,
               intoStage1_CanRun,
               intoStage1_ShouldStop,
               fromStage2_BranchIsTaken,
               fromStage1_PC,
               fromStage1_IR,
               fromStage1_NextPC,
               stage1IsStalled);
//
// Module: FeedbackDRegisterWith1Input
//
// Description:
//      Registers for interface between stage 1 and stage 2
//
// Inputs:
//      clk              <-- clock
//      shouldHold       <-- stage2IsStalled
//      d
//      reset            <-- srcProcessorReset
//      resetValue
//
// Outputs:
//      q
//

FeedbackDRegisterWith1Input #(32, 5, 0, 0) IR2
                                  ( clock,
                                    stage2IsStalled,
                                    fromStage1_IR,
                                    intoStage2_IR,
                                    srcProcessorReset,
                                    32'hF0000000);
FeedbackDRegisterWith1Input #(32, 5, 0, 0) PC2
                                  ( clock,
                                    stage2IsStalled,
                                    fromStage1_PC,
                                    intoStage2_PC,
                                    srcProcessorReset,
                                    32'b0);

FeedbackDRegisterWith1Input #(32, 5, 0, 0) NextPC2
                                  ( clock,
                                    stage2IsStalled,
                                    fromStage1_NextPC,
```

```
                                        intoStage2_NextPC,
                                        srcProcessorReset,
                                        32'b0);

//
// Module: Stage2
//
// Description:
//     Instruction Decode and Operand Read
//
// Inputs from Stage1 to Stage2:
//     inputPC                         <-- intoStage2_PC
//     inputIR                         <-- intoStage2_IR
//     inputNextPC                     <-- intoStage2_NextPC
//
// Outputs from Stage2 to Stage3:
//     outputPC                        --> fromStage2_PC
//     outputIR                        --> fromStage2_IR
//     outputDecodedIR                 --> fromStage2_DecodedIR
//     outputX                         --> fromStage2_X
//     outputY                         --> fromStage2_Y
//     outputMD                        --> fromStage2_MD
//     outputIsRegisterWrite           --> fromStage2_
//                                         IsRegisterWrite
//     outputRa                        --> fromStage2_Ra
//
// Outputs from Stage2 to Stage1 and PC register:
//     outputBranchIsTaken             --> fromStage2_
//                                         BranchIsTaken
//     outputNextPC                    --> fromStage2_NextPC
//
// Output to indicate that Stage2 current sees stop
//    instruction:
//     outputIsStop                    --> fromStage2_IsStop
//
// Interface with Register File:
//     ra                              --> ra
//     contentsRaFromRegisterFile <-- contentsRaFrom
//                                         RegisterFile
//     rb                              --> rb
//     contentsRbFromRegisterFile <-- contentsRbFrom
//                                         RegisterFile
//     rc                              --> rc
//     contentsRcFromRegisterFile <-- contentsRcFrom
//                                         RegisterFile
//
// Interface with Stage3 for forwarding:
//     isRegisterWriteInStage3    <--  isRegisterWrite
//                                         InStage3
```

```
//    raInStage3                       <-- raInStage3
//    contentsRaFromStage3             <-- contentsRaFromStage3
//.   contentsRaFromStage3Ready <-- contentsRaFromStage3
                                           Ready
//
// Interface with Stage4 for forwarding:
//    isRegisterWriteInStage4     <-- isRegisterWrite
                                         InStage4
//    raInStage4                       <-- raInStage4
//    contentsRaFromStage4             <-- contentsRaFromStage4
//    contentsRaFromStage4Ready <-- contentsRaFromStage
                                         4Ready
//
// Interface with Stage5 for forwarding:
//    isRegisterWriteInStage5     <-- isRegisterWrite
                                         InStage5
//    raInStage5                       <-- raInStage5
//    contentsRaFromStage5             <-- contentsRaFromStage5
//    contentsRaFromStage5Ready <-- contentsRaFromStage
                                         5Ready
//
// Output to Stage1 to indicate stall condition:
//    stage2IsStalled                  --> stage2IsStalled
//
// Selectively enable forwarding for experimentation:
//    enableForwarding                 <-- enableForwarding
//

Stage2 stage2 (intoStage2_PC,
               intoStage2_IR,
               intoStage2_NextPC,
               fromStage2_PC,
               fromStage2_IR,
               fromStage2_DecodedIR,
               fromStage2_X,
               fromStage2_Y,
               fromStage2_MD,
               fromStage2_IsRegisterWrite,
               fromStage2_Ra,
               fromStage2_BranchIsTaken,
               fromStage2_NextPC,
               fromStage2_IsStop,
               ra,
               contentsRaFromRegisterFile,
               rb,
               contentsRbFromRegisterFile,
               rc,
               contentsRcFromRegisterFile,
               isRegisterWriteInStage3,
```

```
                    raInStage3,
                    contentsRaFromStage3,
                    contentsRaFromStage3Ready,
                    isRegisterWriteInStage4,
                    raInStage4,
                    contentsRaFromStage4,
                    contentsRaFromStage4Ready,
                    isRegisterWriteInStage5,
                    raInStage5,
                    contentsRaFromStage5,
                    contentsRaFromStage5Ready,
                    stage2IsStalled,
                    enableForwarding);

//
// Module: DRegister
//
// Description:
//     Registers for interface between stage 2 and stage 3
//
// Inputs:
//     clk              <-- clock
//     d
//     reset            <-- srcProcessorReset
//     resetValue
//
// Outputs:
//     q
//
DRegister #(32, 5, 0, 0) PC3 ( clock,
                               fromStage2_PC,
                               intoStage3_PC,
                               srcProcessorReset,
                               32'b0);

DRegister #(32, 5, 0, 0) IR3 ( clock,
                               fromStage2_IR,
                               intoStage3_IR,
                               srcProcessorReset,
                               32'hF0000000);

DRegister #(32, 5, 0, 0) DecodedIR3
                             ( clock,
                               fromStage2_DecodedIR,
                               intoStage3_DecodedIR,
                               srcProcessorReset,
                               32'h40000000);

DRegister #(32, 5, 0, 0) X3 ( clock,
                              fromStage2_X,
```

```
                                        intoStage3_X,
                                        srcProcessorReset,
                                        32'b0);

DRegister #(32, 5, 0, 0) Y3    ( clock,
                                        fromStage2_Y,
                                        intoStage3_Y,
                                        srcProcessorReset,
                                        32'b0);

DRegister #(32, 5, 0, 0) MD3   ( clock,
                                        fromStage2_MD,
                                        intoStage3_MD,
                                        srcProcessorReset,
                                        32'b0);

DRegister #(1, 5, 0, 0) IsRegisterWrite3
                                ( clock,
                                        fromStage2_IsRegisterWrite,
                                        intoStage3_IsRegisterWrite,
                                        srcProcessorReset,
                                        1'b0);

DRegister #(5, 5, 0, 0) Ra3    ( clock,
                                        fromStage2_Ra,
                                        intoStage3_Ra,
                                        srcProcessorReset,
                                        5'b0);

//
// Module: FeedbackDRegisterWith1Input
//
// Description:
//    Registers for interface between stage  1 and stage 2
//
// Inputs:
//    clk                     <-- clock
//    shouldHold              <-- hasDecodedStop
//    d                       <-- fromStage2_IsStop
//    reset                   <-- srcProcessorReset
//    resetValue
//
// Outputs:
//    q                       --> hasDecodedStop
//


FeedbackDRegisterWith1Input #(1, 5, 0, 0) HasDecodedStop
                                ( clock,
                                        hasDecodedStop,
                                        fromStage2_IsStop,
```

```
                                      hasDecodedStop,
                                      srcProcessorReset,
                                      1'b0);

//
// Module: Stage3
//
// Description:
//     ALU operations
//
// Inputs from Stage2:
//     inputPC                <-- intoStage3_PC
//     inputIR                <-- intoStage3_IR
//     inputDecodedIR         <-- intoStage3_DecodedIR
//     inputX                 <-- intoStage3_X
//     inputY                 <-- intoStage3_Y
//     inputMD                <-- intoStage3_MD
//     inputIsRegisterWrite   <-- intoStage3_IsRegister
                                      Write
//     inputRa                <-- intoStage3_Ra
//
// Outputs to Stage3:
//     outputPC               --> fromStage3_PC
//     outputIR               --> fromStage3_IR
//     outputDecodedIR        --> fromStage3_DecodedIR
//     outputZ                --> fromStage3_Z
//     outputMD               --> fromStage3_MD
//     outputIsRegisterWrite  --> fromStage3_IsRegister
                                      Write
//     outputRa               --> fromStage3_Ra
//
// Interface with Stage2 for forwarding:
//     isRegisterWrite        --> isRegisterWriteInStage3
//     ra                     --> raInStage3
//     contentsRa             --> contentsRaFromStage3
//     contentsRaReady        --> contentsRaFromStage3Ready
//

Stage3 stage3 (intoStage3_PC,
               intoStage3_IR,
               intoStage3_DecodedIR,
               intoStage3_X,
               intoStage3_Y,
               intoStage3_MD,
               intoStage3_IsRegisterWrite,
               intoStage3_Ra,
               fromStage3_PC,
               fromStage3_IR,
               fromStage3_DecodedIR,
```

```
                    fromStage3_Z,
                    fromStage3_MD,
                    fromStage3_IsRegisterWrite,
                    fromStage3_Ra,
                    isRegisterWriteInStage3,
                    raInStage3,
                    contentsRaFromStage3,
                    contentsRaFromStage3Ready);

//
// Module: DRegister
//
// Description:
//     Registers for interface between stage 3 and stage 4
//
// Inputs:
//     clk                       <-- clock
//     d
//     reset                     <-- srcProcessorReset
//     resetValue
//
// Outputs:
//     q
//


DRegister #(32, 5, 0, 0) PC4 ( clock,
                               fromStage3_PC,
                               intoStage4_PC,
                               srcProcessorReset,
                               32'b0);

DRegister #(32, 5, 0, 0) IR4 ( clock,
                               fromStage3_IR,
                               intoStage4_IR,
                               srcProcessorReset,
                               32'hF0000000);

DRegister #(32, 5, 0, 0) DecodedIR4
                             ( clock,
                               fromStage3_DecodedIR,
                               intoStage4_DecodedIR,
                               srcProcessorReset,
                               32'h40000000);

DRegister #(32, 5, 0, 0) Z4  ( clock,
                               fromStage3_Z,
                               intoStage4_Z,
                               srcProcessorReset,
                               32'b0);
```

```
DRegister #(32, 5, 0, 0) MD4 ( clock,
                               fromStage3_MD,
                               intoStage4_MD,
                               srcProcessorReset,
                               32'b0);


DRegister #(1, 5, 0, 0) IsRegisterWrite4
                             ( clock,
                               fromStage3_IsRegisterWrite,
                               intoStage4_IsRegisterWrite,
                               srcProcessorReset,
                               1'b0);


DRegister #(5, 5, 0, 0) Ra4 ( clock,
                               fromStage3_Ra,
                               intoStage4_Ra,
                               srcProcessorReset,
                               5'b0);



//
// Module: Stage4
//
// Description:
// Memory Access
//
// Inputs from Stage3:
//    inputPC               <-- intoStage4_PC
//    inputIR               <-- intoStage4_IR
//    inputDecodedIR        <-- intoStage4_DecodedIR
//    inputZ                <-- intoStage4_Z
//    inputMD               <-- intoStage4_MD
//    inputIsRegisterWrite  <-- intoStage4_IsRegisterWrite
//    inputRa               <-- intoStage4_Ra
//
// Clock phase for qualifying write strobe:
//    qualifierClock        <-- clock
//
// Outputs to Stage5:
//    outputPC              --> fromStage4_PC
//    outputIR              --> fromStage4_IR
//    outputDecodedIR       --> fromStage4_DecodedIR
//    outputZ               --> fromStage4_Z
//    outputIsRegisterWrite --> fromStage4_
//                              IsRegisterWrite
//    outputRa              --> fromStage4_Ra
//
// Interface with Stage2 for forwarding:
//    isRegisterWrite       --> isRegisterWriteInStage4
```

```
//     ra                        --> raInStage4
//     contentsRa                --> contentsRaFromStage4
//     contentsRaReady           --> contentsRaFromStage4Ready
//
// Interface with data memory:
//     toDataMemory_Address       --> toDataMemory_Address
//     toDataMemory_Data          --> toDataMemory_Data
//     toDataMemory_WriteStrobe  --> toDataMemory_Write
                                      Strobe
//     fromDataMemory_Data        <-- fromDataMemory_Data
//

Stage4 stage4 (intoStage4_PC,
               intoStage4_IR,
               intoStage4_DecodedIR,
               intoStage4_Z,
               intoStage4_MD,
               intoStage4_IsRegisterWrite,
               intoStage4_Ra,
               clock,
               fromStage4_PC,
               fromStage4_IR,
               fromStage4_DecodedIR,
               fromStage4_Z,
               fromStage4_IsRegisterWrite,
               fromStage4_Ra,
               isRegisterWriteInStage4,
               raInStage4,
               contentsRaFromStage4,
               contentsRaFromStage4Ready,
               toDataMemory_Address,
               toDataMemory_Data,
               toDataMemory_WriteStrobe,
               fromDataMemory_Data);
//
// Module: DRegister
//
// Description:
//     Registers for interface between stage 4 and stage 5
//
// Inputs:
//     clk           <-- clock
//     d
//     reset         <-- srcProcessorReset
//     resetValue
//
// Outputs:
//     q
//
```

```
DRegister #(32, 5, 0, 0) IR5 ( clock,
                                fromStage4_IR,
                                intoStage5_IR,
                                srcProcessorReset,
                                32'hF0000000);

DRegister #(32, 5, 0, 0) PC5 ( clock,
                                fromStage4_PC,
                                intoStage5_PC,
                                srcProcessorReset,
                                32'b0);


DRegister #(32, 5, 0, 0) DecodedIR5
                          ( clock,
                            fromStage4_DecodedIR,
                            intoStage5_DecodedIR,
                            srcProcessorReset,
                            32'h40000000);

DRegister #(32, 5, 0, 0) Z5  ( clock,
                               fromStage4_Z,
                               intoStage5_Z,
                               srcProcessorReset,
                               32'b0);


DRegister #(1, 5, 0, 0) IsRegisterWrite5
                          ( clock,
                            fromStage4_IsRegisterWrite,
                            intoStage5_IsRegisterWrite,
                            srcProcessorReset,
                            1'b0);


DRegister #(5, 5, 0, 0) Ra5  ( clock,
                               fromStage4_Ra,
                               intoStage5_Ra,
                               srcProcessorReset,
                               5'b0);


//
// Module: Stage5
//
// Description:
//     Register Write
//
// Inputs from Stage4:
//     inputPC              <-- intoStage5_PC
//     inputIR              <-- intoStage5_IR
//     inputDecodedIR       <-- intoStage5_DecodedIR
```

```
//      inputZ                  <-- intoStage5_Z
//      inputIsRegisterWrite <-- intoStage5_IsRegisterWrite
//      inputRa                 <-- intoStage5_Ra
//
// Clock phase for qualifying write strobe:
//      qualifierClock          <-- clock
//
// Outputs from Stage5:
//      outputIsStop            --> fromStage5_IsStop
//
// Interface with Stage2 for forwarding:
//      isRegisterWrite         --> isRegisterWriteInStage5
//      ra                      --> raInStage5
//      contentsRa              --> contentsRaFromStage5
//      contentsRaReady         --> contentsRaFromStage5Ready
//
// Interface with data memory:
//      toRegisterFile_Address     --> toRegisterFile_
//                                     Address
//      toRegisterFile_Data        --> toRegisterFile_Data
//      toRegisterFile_WriteStrobe --> toRegisterFile_
//                                     WriteStrobe
//

Stage5 stage5 ( intoStage5_PC,
                intoStage5_IR,
                intoStage5_DecodedIR,
                intoStage5_Z,
                intoStage5_IsRegisterWrite,
                intoStage5_Ra,
                clock,
                fromStage5_IsStop,
                isRegisterWriteInStage5,
                raInStage5,
                contentsRaFromStage5,
                contentsRaFromStage5Ready,
                toRegisterFile_Address,
                toRegisterFile_Data,
                toRegisterFile_WriteStrobe);

FeedbackDRegisterWith1Input #(1, 5, 0, 0) HasExecute Stop(
                clock,
                hasExecutedStop,
                fromStage5_IsStop,
                hasExecutedStop,
                srcProcessorReset,
                1'b0);
endmodule
```

Bibliography

MARK ALEXANDER. Power distribution system (PDS) design: using bypass/decoupling capacitors. Xilinx XAPP623, San Jose, CA 95124 February 2005.

PETER ALFKE AND CLIFFORD E. CUMMINGS. Simulation and synthesis techniques for asynchronous FIFO design with asynchronous pointer comparisons. SNUG 2002 (Synopsys Users Group), San Jose, CA April 2002.

KEN CHAPMAN. Get smart about reset (think local, not global). Xilinx TechXClusives, San Jose, CA 95124 October 2001.

KEN COFFMAN. *Real World FPGA Design with Verilog*. Prentice Hall, Upper Saddle River, NJ, 2000.

CLIFFORD E. CUMMINGS. Full case parallel case, the evil twins of verilog synthesis. SNUG 1999 (Synopsys Users Group) Boston MA.

CLIFFORD E. CUMMINGS. Synthesis and scripting techniques for designing multi-asynchronous clock designs. SNUG 2001 San Jose, CA (Synopsys Users Group), March 2001.

CLIFFORD E. CUMMINGS. New Verilog-2001 techniques for creating parameterized models (or down with define and death of a defparam!). HDLCON 2002, San Jose, CA May 2002.

CLIFFORD E. CUMMINGS, STEVE GOLSON, AND DON MILLS. Asynchronous & synchronous reset design techniques — part deux. SNUG 2003 (Synopsys Users Group), Boston, MA September 2003.

WILLIAM J. DALLY AND JOHN W. POULTON. *Digital Systems Engineering*. Cambridge University Press, Cambridge, UK, 1998.

VINCENT P. HEURING AND HARRY F. JORDAN. *Computer Systems Design and Architecture*. Addison Wesley Longmann, Menlo Park, CA, 1997.

PHILIPPE GARRAULT AND BRIAN PHILOFSKY. HDL coding practices to accelerate design performance. Xilinx White Paper WP231, San Jose, CA 95124 January 2006.

HOWARD W. JOHNSON AND MARTIN GRAHAM. *High-Speed Digital Design: A Handbook of Black Magic*. Prentice Hall, Upper Saddle River, NJ, 1992.

The Institute of Electrical and Electronics Engineers (IEEE). IEEE Standard for Binary Floating-Point Arithmetic. IEEE Standards Board, New York, NY March 1985.

National Institute of Standards and Technology (NIST). Advanced Encryption Standard (AES). Federal Information Processing Standards Publication 197, Gaithersburg, MD 20899 November 2001.

National Institute of Standards and Technology (NIST). Secure Hash Standard (SHA). Federal Information Processing Standards Publication 180-2, Gaithersburg, MD 20899 2001

SAMIR PALNITKAR. *Verilog HDL, A Guide to Digital Design and Synthesis*. Prentice Hall, Upper Saddle River, NJ, 1996.

Advanced FPGA Design. By Steve Kilts
Copyright © 2007 John Wiley & Sons, Inc.

BOAZ PORAT. *A Course in Digital Signal Processing.* John Wiley & Sons, New York, 1997.

Synplicity Inc. Fast timing closure on FPGA designs using graph-based physical synthesis. Synplicity White papers, Sunnyvale, CA 94086 September 2005.

Index

Advanced FPGA Design. By Steve Kilts
Copyright © 2007 John Wiley & Sons, Inc.

321